Advances in Electric Power Engineering

Editor
Hajar Bagheri Tolabi

FIRST EDITION

LAXMI BOOK PUBLICATION
258/34, Raviwar Peth,
Solapur-413005
Cell: +91 9595359435

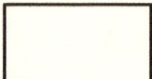

Rs: **/-**

"Advances in Electric Power Engineering"

Hajar Bagheri Tolabi

© 2015 by Laxmi Book Publication, Solapur

ISBN-

Published by,

Laxmi Book Publication,
258/34, Raviwar Peth,
Solapur, Maharashtra, India.

Contact No. : +91 9595 359 435
Website : http://www.isrj.org
Email ID : ayisrj@yahoo.in

Preface

In the 21st century, electric power engineering is going green and smart. In this century, the increasing search for the efficiency, the computational continuous improvement and the development of new effective mathematical methods are three impelling forces for the utilization of optimization in electric power systems.

Nowadays, it is unlikely to find an electric company that does not use optimization methods. This kind of processes is utilized in both planning and operation calculations for the generation, transmission and distribution areas of power systems. Electrical engineers face these new operational methods, in some cases without the adequate preparation. This book aims to include some of the present and foreseen applications of the optimization in electric power systems, explained by main experts in the field. Furthermore, this book may serve as state-of-the-art for undergraduate and graduate students worldwide.

As can be seen, the chapters in each section maintain their own thematic continuity and at the same time have significant overlaps with chapters in other sections as well. Therefore, one may read the book in its entirety or focus on individual chapters. Due to its broad scope, this will be an ideal resource for students in advanced graduate-level courses and special topics in power systems. It will also interest utility engineers who seek an intuitive understanding of the emerging applications of optimization methods in power systems.

Hajar Bagheri Tolabi

Acknowledgement

I would like to put on record my heartfelt gratitude to my beloved teachers Dr. M. H. Moradi (Bu Ali Sina University, Iran), and Dr M. H. Ali (University of Memphis, USA) for contributing their valuable works in the completion of this book.

I am grateful to Dr. Ashok Yakkaldevi for excellent secretarial assistance. Finally, I thank my parents and my husband for their patience with me.

About Editor

 Hajar Bagheri Tolabi is a Ph.D. Research Scholar and a Faculty member at Khorramabad Branch, Islamic Azad University, Khorramabad, Iran. She has five years of teaching experience including four years of experience in the field of Research. The author has a special inclination towards academic research. She has published one academic book and over 20 research papers in national and in international journals. She has presented several papers in seminars/conferences at national and international levels. The author also serves as a reviewer in many reputed Journals. Recently in the year 2014 he has been awarded as one of the best selected researchers at Lorestan State, Iran.

Her area of interest includes renewable energy, distribution systems and Optimization

List of Authors

Hajar Bagheri Tolabi
Faculty of engineering,
 Islamic Azad University,
Khorramabad Branch,
Khorramabad, Iran

Mohammad Faridun Naim Tajuddin,
School of Electrical System Engineering,
Universiti Malaysia Perlis (UniMAP),
Pauh Putra, Perlis,
Malaysia-02000

Mohammad H. Moradi
Faculty of Engineering,
Department of Electrical Engineering,
Bu-Ali Sina University,
Hamedan, Iran

Sheeraz Kirmani
Department of Electrical Engineering
Jamia Millia Islamia,
New Delhi-110025

M. Rizwan
Department of Electrical Engineering, Delhi Technological University, Delhi, India

Rahil Hossein
Department of Computer Engineering,
Shahr-e-Qods Branch,
Islamic Azad University,
Tehran, Iran

Mahmoud Reza Shakarami
Faculty of engineering,
Lorestan University,
Khorramabad, Iran

Shahrin Md Ayob
Faculty of Electrical Engineering,
Universiti Teknologi Malaysia (UTM), UTM Skudai,
Johor, Malaysia-81310

Majid Jamil
Department of Electrical
Engineering,
Jamia Millia Islamia,
New Delhi, India

S M. Reza Tousi
Faculty of Engineering,

Department of Electrical
Engineering,
Bu-Ali Sina University,
Hamedan, Iran

Zainal Salam
Faculty of Electrical Engineering,
Universiti Teknologi Malaysia
(UTM),
UTM Skudai,
Johor, Malaysia-81310

A. Hatami
Faculty of Engineering,
Department of Electrical
Engineering,
Bu-Ali Sina University,
Hamedan, Iran

A. Kohansali
Faculty of Engineering,
Department of Electrical
Engineering,
Bu-Ali Sina University,
Hamedan, Iran

Priyanka Chaudhary
Technological University,
Delhi, India

Esmaeel Rok Rok
Faculty of engineering,
Lorestan University,
Khorramabad, Iran

Sagnika Ghosh
Department of Electrical and
Computer Engineering, University
of Memphis, USA

Mohd Hasan Ali
Department of Electrical and
Computer Engineering, University
of Memphis, USA

Contents

Chapter 1
A New Energy Management Strategy for Smart Buildings
Based On Advance Metering

Mohammad H. Moradi, S M. Reza Tousi, A. Kohansali and

A.Hatami

Abstract-

In this Chapter, a method for optimizing 24-hour scheduling of several neighboring buildings equipped with independent DERs is proposed. The chosen buildings have different electrical energy consumption and load profiles, and the buildings' thermal demands are modeled more precisely. The data provided by smart meters enable buildings to participate in demand response programs and the communication infrastructures among the buildings make the drawing of bilateral contracts possible. Three strategies named independent, cooperative, and centralized operations of buildings are investigated. The outcome of optimizations in these scenarios shows that the daily operational cost can be considerably reduced through reciprocation / cooperation among the buildings.

Nomenclature

Variables	
h	Hours of day (1, 2, … ,24)
$EP(h)$	Main grid energy price at hour h
$P_{grid}(h)$	exchange power with main grid at hour h
$G_{C\&A}(h)$	consumed natural gas by CHP and auxiliary boiler
$IEP(h)$	Inter agents energy price at hour h
$P_I(h)$	Inter building exchanged power at hour h
$T_s(h)$	Water storage temperature at hour h
$V_{cold}(h)$	Replaced cold water volume at hour h
$H_{CHP}(h)$	CHP thermal output power at hour h
$H_{aux}(h)$	Auxiliary boiler output power at hour h
$H_{air}(h)$	Thermal power needed to set building temperature at hour h
$T_{out}(h)$	Outdoor temperature at hour h
$P_{chp}(h)$	CHP electrical output power at hour h
$P_{batt}(h)$	Charged or discharged power of battery at hour h
$P_{PV}(h)$	PV cells output power at hour h
$L_{nshf}(h)$	Total Nonshiftable loads at hour h
$L_{shf}^{n}(h)$	Power consumption of nth shiftable load at hour h
$c_{aux}(h)$	Auxiliary boiler state at hour h, 1:On, 0:OFF
$c_{chp}(h)$	CHP state at hour h, 1:On, 0:OFF
$a_n(h)$	n^{th} shiftable load state at hour h, 1:On, 0:OFF
$SOC(h)$	Battery state of charge at hour h

2

$dch(h)$	Battery discharging state at hour h, 1: discharging, 0: otherwise
$ch(h)$	Battery charging state at hour h, 1: charging, 0: otherwise
$P_{batt}^{ch}(h), P_{batt}^{dch}(h)$	Battery's charging and discharging power at hour h
Constants	
h_n	Available hours of operation of shiftable loads
GP	Natural gas price
T_s^c	Replaced cold water temperature
$P_{chp}^{max}, P_{chp}^{min}$	CHP output power maximum and minimum
T_s^{max}, T_s^{min}	Hot water storage temperature maximum and minimum
$T_{in}^{max}, T_{in}^{min}$	Indoor temperature maximum and minimum
$H_{aux}^{max}, H_{aux}^{min}$	Auxiliary boiler output power maximum and minimum
V_s	Hot water storage volume
R	Thermal resistance of buildings' shell
C_{gas}	Natural gas consumption rate for producing 1kWh energy
N	Number of shiftable loads
η_{th}, η_{elec}	CHP thermal and electrical efficiency
η_{aux}	Auxiliary boiler efficiency
P_{rr}	CHP ramp rate
HOP_n	n^{th} shiftable load hours of operation
E_{total}^n	Total energy consumption of nth shiftable load
η_{dch}, η_{ch}	Charge and discharge efficiency

SOC_{max}, SOC_{min}	Battery state of charge maximum and minimum
Indices	
" C "	Indices stand for commercial building
" R "	Indices stand for residential building
" H "	Indices stand for Hospital building

1.1 Introduction

The liberalization of electricity market and necessity of reducing greenhouse gas emission and world energy crisis have led to development of renewable energy resources and alternative generation systems with higher efficiency [1], [2]. Since establishment and operation of large power plants are expensive and low efficient, Micro-grids and smart gird technologies are known as solutions of modern power grid [3]. Micro-grid is a group of loads and distributed energy resources (DER) as a single controllable system. A micro grid can utilize Smart tools such as smart meters and communicational infrastructures to use the available energy more efficiently by performing energy management programs such as demand side management (DSM).[4],[5].

DSM programs mostly refer to energy efficiency programs, demand response programs, and residential or commercial load management [6]. The purpose of residential load management programs is one or both of reducing consumption and shifting consumption [8]. DSM programs are done according to interact and exchange information with the control center of energy provider company. In [9] an incentive-based energy consumption scheduling scheme has been proposed in case where a single energy source is shared by several customers and the customers are equipped with smart meters that are connected to power line and communication network simultaneously. In [10] the scheduling problem of building energy supplies is considered and the objective function is to minimize

4

the total cost of electricity and natural gas for a building facilitated by Microgird while satisfying the energy balance and complicated operating constraints of individual energy supply equipment and devices. Economical scheduling of a microgrid in an isolated load area is performed by mixed integer linear programing (MILP) in [12]. It has been shown in [13] that the separation of shiftable and nonshiftable loads would lead to a better microgrid participation in implementing demand response programs. It has also showed that flexible thermal load's would have good effects on micro-CHP unit operation. In many studies thermal loads are considered as a single profile without the separation of desired hot water and indoor temperature in each time interval [14]. In [15] an equipped building with smart meters and a micro-CHP is considered and the thermal loads are separated and modeled exactly, so the flexible thermal loads and shiftable electrical loads are coordinated and then the scheduling problem has been solved. An optimal scheduling of smart homes' energy consumption has been performed by using MILP in [16] and the household tasks are scheduled based on real-time electricity pricing, electricity task time window and forecasted renewable energy output.

Another considerable point in operating of microgrids is energy management and optimization of them. The importance of planning and energy management becomes more marked when the recent studies shows that 20%-30% of building energy consumption can be saved through optimized operation and management without changing the building structure [17]. In Dec. 2009 in Copenhagen, by the Major Economies Forum (MEF), a technology action plan for smart grids announced which in it "active demand response" and "integration with smart home" are considered as the first items on the smart grids technology fact sheet [18]. Smart meters with bidirectional communication capacity are the most important instruments in modern smart grids. These advanced meters are capable of measurement and identification of power consumption electronically and can

5

communicate this data to another device [19]. One of the types of smart meters is the power strip type smart meter (SMPT) which have several sensors for monitoring and controlling power [15], [20].

In this paper, the effects of power transaction capability and data exchange between three neighbor buildings on operating cost and resources planning are studied. It's assumed that the buildings are equipped by smart grid technology and the data is provided by smart meters showing the energy consumption and obtaining a load profile of each building. Smart buildings are assumed to have different usages, residential, commercial, and hospital so they would have different load profile. Buildings operation and minimizing energy costs in some scenarios has performed by MILP model, and by ε-Constraint model in another scenario. In this study, DERs such as CHP units are used in order to produce heat and electrical energy and some electrical loads are shift able. The scheduling is based on a TOU pricing tariffs.

The rest of this paper is organized as follow. The problem formulation, including objective functions and corresponding constraints are presented in section II. Then section III explains buildings operation scenarios. The simulation results are illustrated in section IV. The discussion is in section V and the conclusions are provided in section VI.

1.2 Problem Formulation

The main objective of this study is to investigate the effects of both the transaction capability and data exchange among buildings on operation costs. The scheduling algorithm of buildings is so devised that the decision making unit acquires its needed information from forecasting units or smart meters, and, using the same information, it solves the optimal planning by MILP method. This objective can easily be obtained by programming a digital controller using the Advanced Metering Infrastructure (AMI). AMI system contains units including data processing, electrical and thermal load forecasting, shift able load identifying units, and the units updating set points for CHP units.

1.2.1 Objective Function

The objective for each building is to minimize the cost of its operation. Each building's cost function has three major components, which include: cost of traded energy with the main grid, cost of consumed natural gas in CHP unit and auxiliary boiler, and cost of traded energy with neighbor buildings. It is obvious that when the buildings are operating independently the third part of building objective function is zero. It should be noted that the operation scenarios will be explained in section III completely. The equation (1) shows the generalized objective function of buildings.

$$J = \sum_{h=1}^{24} EP(h).P_{grid}(h) + GP.G_{C\&A}(h) + IEP(h).P_l(h) \tag{1}$$

In this study the price of natural gas is constant during a day. Positive and negative values for $P_{grid}(h)$ represent buying and selling energy from/to grid respectively, and purchasing and selling prices are not necessarily the same at hour h.

1.2.2 Thermal Load Modeling

In this paper, the thermal loads of a building are divided to hot water and heating with a range of variation. Flexibility of these loads cause that the electrical loads and thermal loads would be coordinated with each other. It's assumed that the hot water storage is always full and the used hot water is replaced by cold water. The injected heat from CHP unit or auxiliary boiler and using hot water will change the storage temperature. Equation (2) shows temperature of hot water storage at hour $h+1$ [15]:

$$T_s(h+1) = \frac{V_{cold}(h).(T_s^c - T_s(h)) + V_s.T_s(h)}{V_s} + \frac{H_{CHP}(h) + H_{aux}(h) - H_{air}(h)}{V_s.C_w} \tag{2}$$

In (2), the first term shows the effect of cold water replacement in storage and the second term shows the effect of CHP unit and auxiliary boiler units' heat generation. By solving equation (2) and keeping the storage temperature at hour *h+1*in a desired range (according to constraint in 2.2.2), the amount of need for heat generation will be figured out. The shortage of heat generation by CHP unit can be compensated by auxiliary boiler whose existence can help to coordinate electrical and thermal loads thus ending up to the reduction of gas consumption ($G_{C\&A}$).

According to presented thermal model for building in [21] the indoor temperature of building is as follow:

$$T_{in}(h+1) = T_{in}(h).e^{\frac{-1}{\tau}} + (R.H_{air}(h) + T_{out}(h)).\left(1 - e^{\frac{-1}{\tau}}\right)$$
(3)

Here $\tau = R.C_{air}$ is thermal time constant.

1.2.3 Shiftable Electrical Loads

Shiftable electrical loads are those that can be utilized under predefined time intervals and definite hours of operation. But for the purpose of cost reduction and contribution in the demand side management programs, the decision concerning their exact operation time is taken by building control center. The two parameters of shiftable loads are their hour of operation and total energy consumption which provided by smart meters in the building. By knowing these two parameters, controlling and scheduling of shiftable loads would be changed. Therefore, the power trade with grid (P_{grid}) and neighbors (

P_l) will be changed and shifted to optimize the cost function. Also, the constraints related to these loads are presented in section 2.2.6.

1.2.4   Natural Gas Consumption Calculation

The amount of natural gas that has been consumed by CHP unit and auxiliary boiler is as follow:

$$G_{C\&A}(h) = (P_{chp}(h)/\eta_{elec} + H_{aux}(h)/\eta_{aux}).C_{gas} \qquad (4)$$

1.3. Constraints

1.3.1 Electrical Power Balance Constraint

This constraint guarantees that the energy generation and demand are equal at each hour. In this study the electrical resources are include CHP units, main grid, and in one building PV cells. There is an auxiliary boiler to meet the thermal demand in case of lack of CHP heat generation. The loads are separated into to shiftable and nonshiftable loads and there are batteries as electrical energy storage. The electrical power balance in general state is:

$$P_{grid}(h) - P_{batt}(h) + P_{chp}(h) + P_{PV}(h) = L_{nshf}(h) + \sum_{N}^{n=1} L_{shf}^{n}(h) \qquad (5)$$

1.3.2 Thermal Constraints

As mentioned before in this study it's been assumed that thermal loads are flexible. It means that the hot water storage temperature and indoor temperature can vary along defined boundaries without any impact on human comfort.

$$T_s^{\min} \leq T_s(h) \leq T_s^{\max}$$
$$T_{in}^{\min} \leq T_{in}(h) \leq T_{in}^{\max} \qquad (6)$$

1.3.3 Power Limits of CHP Unit and Auxiliary Boiler:

$$c_{chp}(h).P_{chp}^{min} \leq P_{chp}(h) \leq c_{chp}(h).P_{chp}^{max}$$

$$c_{chp}(h).H_{chp}^{min} \leq H_{chp}(h) \leq c_{chp}(h).H_{chp}^{max}$$

$$c_{aux}(h).H_{aux}^{min} \leq H_{aux}(h) \leq c_{aux}(h).H_{aux}^{max} \quad (7)$$

1.3.4 CHP unit Thermal and Electrical Efficiency:

$$P_{chp}(h) = H_{chp}(h).\frac{\eta_{elec}}{\eta_{th}} \quad (8)$$

1.3.5 CHP Unit Output Ramp Rate

$$-P_{rr} \leq P_{chp}(h) - P_{chp}(h-1) \leq P_{rr}$$

$$-\frac{\eta_{th}}{\eta_{elec}}.P_{rr} \leq H_{chp}(h) - H_{chp}(h-1) \leq \frac{\eta_{th}}{\eta_{elec}}.P_{rr} \quad (9)$$

1.3.6 Constraint of shiftable loads

As mentioned in section 2.1.2 the amount of total consumed energy of each load (E_{total}^n) and the its hours of operation (HOP_n), are important factors for an optimal scheduling. These numbers are measured or predicted by buildings' smart meters, or should be registered as default values in the control and decision making unit. Hourly consumed power of each shiftable load is calculated as:

$$L_{shf}^n(h) = \frac{E_{total}^n}{HOP_n}.a_n(h)\Big|_{\forall h \in h_n}$$

$$\sum_{h \in h_n} L_{shf}^n(h) = E_{total}^n \quad (10)$$

Another significant note about shiftable loads is that their operation duration should be continuous and without interruption until they complete their corresponding task.

$$a_n(h) + a_n(h+1) + \ldots + a_n(h + HOP_n - 1) \geq$$
$$(HOP_n - 1).(a_n(h-1) - a_n(h-2)) \quad , \quad \forall h \in h_n \qquad (11)$$

Also there are some loads with dependent task in a building, such as clothe washer and dryer. Thus in the optimal planning they should be scheduled according to their dependency. If shiftable load m should be operated after shiftable load n then:

$$a_m(h + HOP_n) + a_m(h+1 + HOP_n) + \ldots + a_m(h + HOP_n + HOP_m) \geq$$
$$(HOP_m - 1).(a_n(h-1) - a_n(h-2)) \quad , \quad \forall h \in h_n \qquad (12)$$

1.3.7 Battery Operation constraints

$$0 \leq \frac{P_{batt}^{ch}(h)}{\eta_{ch}} \leq ch(h).P_{ch}^{max}$$

$$0 \leq P_{batt}^{dch}(h).\eta_{dch} \leq dch(h).P_{dch}^{max}$$

$$ch(h) + dch(h) \leq 1$$

$$P_{batt}(h) = \frac{P_{batt}^{ch}(h)}{\eta_{ch}} - P_{batt}^{dch}(h).\eta_{dch}$$

$$SOC(h+1) = SOC(h) + \frac{P_{batt}^{ch}(h) - P_{batt}^{dch}(h)}{E_{batt}}$$

$$SOC_{min} \leq SOC(h) \leq SOC_{max} \qquad (13)$$

1.4 Operation Scenarios

In this section three scenarios are considered to find the optimum solution of equation 1. The scenarios are as follow:

11

1.4.1 Independent decision making strategy

The first approach is independent and autonomous management scheme which does not need any data network connectivity. Each building is minimizing its own objective function and cost without any access to neighbors' information. It is also the most common case in reality and has been used in all previous studies. In this strategy, there are three separated objective functions and the buildings are assumed to be supplied with their own generation units or utility grid (14)

$$\begin{cases} \min_{A_C} Z_C \\ \min_{A_R} Z_R \\ \min_{A_H} Z_H \end{cases} \Bigg| \quad A_C \cap A_R \cap A_H = \varnothing \tag{14}$$

Where A is set point vectors of all power resources and C,R, and H indexes stand for Commercial, Residential, and Hospital building

1.4.2 Independent decision making strategy with cooperation among neighbors

In this scenario, the scheduling would be performed under assumption of presence of a data exchange network. This ability provides buildings more information to decide more efficiently on their 24-hour power scheduling in hope to reduce their costs. The buildings are also able to make bilateral contracts for purchasing or selling power to each other.

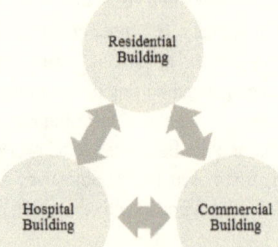

Fig 1Architecture of Cooperative decision

The utility grid provides the path of transmitting electrical energy among buildings and charges them a little for providing this service. Since the price of local generation by DER is cheaper than purchasing energy from utility grid thus some buildings would be interested to buy their power demand from their neighbors instead of only grid utility. Fig1 shows the architecture of the independent decision making strategy with cooperation among neighbors. In this case, there are three dependent optimization problems which must be solved based not only on local information but also the data from neighbors (15)

$$\begin{cases} \min_{A_C} Z_C \\ \min_{A_R} Z_R \big| A_C \cap A_R \cap A_H \neq \emptyset \\ \min_{A_H} Z_H \end{cases}$$

$$(15)$$

1.4.3 Centralized decision making strategy (Centralized control)

In this scenario, the scheduling is performed in a central decision making unit. It is assumed that a central controller is available. This unit would calculate proper set points of all supplies based on a unique objective function (16) which considers minimization of costs of all buildings simultaneously. So, in this way the summation of all costs of buildings will be minimized and there is a probability of non-optimum solution from the point of view of a specific building. The buildings have no control over their generating units and the set-points would be dictated from central control unit.

$$\min_{(A_C \cup A_R \cup A_H)} (Z_C + Z_R + Z_H) \tag{16}$$

1.5 Simulation Resualts

As stated before, in this paper three hospital, commercial and residential buildings have been considered for power scheduling problem. These buildings have different load profiles during the day, so the proposed strategy can be checked easily. It has been assumed that the buildings are equipped with smart meters and proper communicational infrastructures. By these assumptions shiftable loads from non shiftable loads and desired hot water from required energy for controlling indoor temperature could be separable.

1. The first building considered is a 20-apartment residential complex in which a CHP unit has been used as DER. An auxiliary boiler is used to generate thermal energy in case CHP unit heat production is found inadequate. A battery energy storage unit has also been used to improve system reliability and help to reduce micro grid operation cost. Besides, there are some home appliances in each apartment which are controllable, shiftable, and have the capability to affect demand scheduling program. Of course the shiftability is carried out based on above mentioned limitations.

2. The next construction considered for study is a hospital building. Energy generating sources such as CHP, battery storage unit, and auxiliary boiler units in this building are generally similar in type to, while different in capacity, from those in the residential one. There are sterilizing electrical devices in the hospital which can be considered as shiftable loads provided that they have performed

14

their task during the previous night, so that the sterilized instruments are ready the next morning.

3. The third building is a commercial one. Besides the sources mention in the two former buildings, there is in this building a PV generating unit which helps to bring about a considerable reduction in its operation cost.

The buildings' electrical and thermal demands are shown in Fig. 2, Fig. 3, andThe price of the purchased natural gas in this study is quoted 0.32 $/m³, and the TOU tariffs are as in Table 1 [15], and the Table 2 shows sources capacity and assumed parameters for solving .

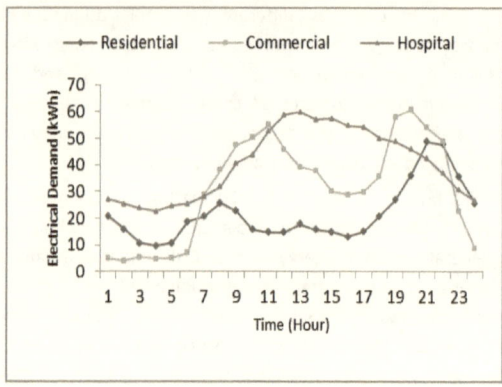

Fig 2 Buildings' Electrical Demand

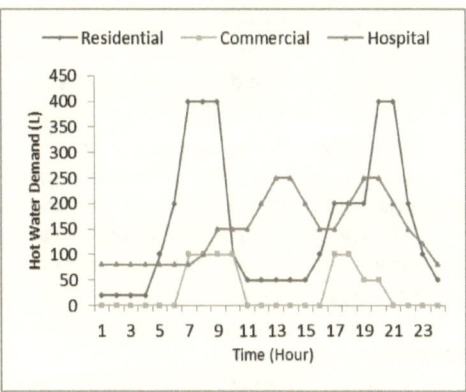

Fig 3 Buildings' Desired Hot Water

Table 1 Electricity Tariff

HOUR	Price ($/kWh)	HOUR	Price ($/kWh)
1	0.062	13	0.092
2	0.062	14	0.092
3	0.062	15	0.092
4	0.062	16	0.092
5	0.062	17	0.092
6	0.062	18	0.108
7	0.062	19	0.108
8	0.108	20	0.062
9	0.108	21	0.062
10	0.108	22	0.062
11	0.108	23	0.062
12	0.092	24	0.062

The energy consumption and hours of operation of shiftable home appliances and hospital equipment are in Table 3. Since all home appliances are not being used in a day a coefficient has been considered. The price of trading energy between buildings a assumed to be 9.1 (¢/kWh), this price is a little less than the average of purchasing price from grid. In this study the optimizations have been perform by MILP method and ε-Constraint in GAMS.

The operation costs for each building in three scenarios are shown in Table 4. In first scenario, independent decision making strategy, each building's operation cost is independent from two other and is in its minimum state. In [15], [16] it has be shown that the presence of shiftable appliances can lead to cost reduction, so in these cases they would have similar positive effects of operation cost. According to the optimal scheduling performed for residential building in first scenario only five of twenty washing machines operate in hour 10th and the others are distributed in the day long. This condition, operation time distribution along day, is similar for the other shiftable loads. Figs 4-6 show the most of buildings energy demand has been prepared by CHP and the presence of auxiliary boiler is necessary for heat production in case of the CHP produced thermal energy is not sufficient.

Table 2 Sources Capacity and Assumed Parameters

	Commercial	Residential	Hospital	UNIT
$T_{in}^{max}, T_{in}^{min}$	19–24	23–27	20–25	°C
T_s^{min}, T_s^{max}	50–60	60–80	60–75	°C
E_{batt}	20	30	40	kWh
$P_{chp}^{max}, P_{chp}^{min}$	3–30	5–40	5.5–55	kWh
η_{th}, η_{elec}	0.3–0.5	0.3–0.5	0.3–0.5	%
η_{aux}	0.98	0.98	0.98	%
η_{dch}, η_{ch}	0.98–0.85	0.98–0.85	0.98–0.85	%

17

SOC_{min}, SOC_{max}	0.3–1	0.3–1	0.3–1	(p.u)
H_{aux}^{max} $\cdot H_{aux}^{min}$	5–50	3–30	4–40	kWh
C_w	4.18	4.18	4.18	$kWh / {}^{\circ}C$
C_{air}	0.525	0.525	0.525	$kWh / {}^{\circ}C$
P_{rr}	10	10	10	kW/h
C_{gas}	0.0925	0.0925	0.0925	kJ/Nm³

Table 3 Shiftable Loads Data

	Total Energy Consumption	Hours of Operation
Washing machine	1	2
Dish washer	1.3	1
Vacuum cleaner	1.1	1
Dryer	2.7	3
Iron	1.3	1
Sterilization equipment	20	4

Table 3 Operation Costs

	Independent decision making strategy	Cooperative decision making strategy	Centralized decision making strategy
Residential	64.44	59.54	62.25
Hospital	112.13	108.10	90.54
Commercial	34.94	28.35	30.58
TOTAL	211.51	195.99	183.37

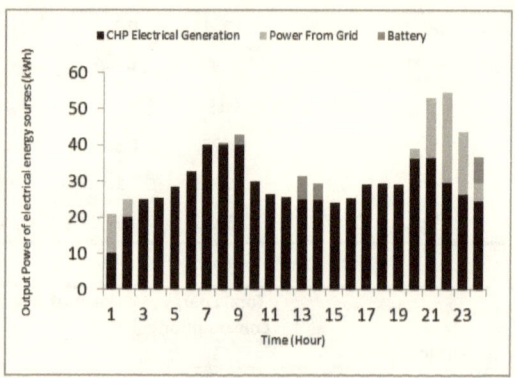

Fig 4 Composition of Electrical Energy Supplies for Residential Building in First Scenario

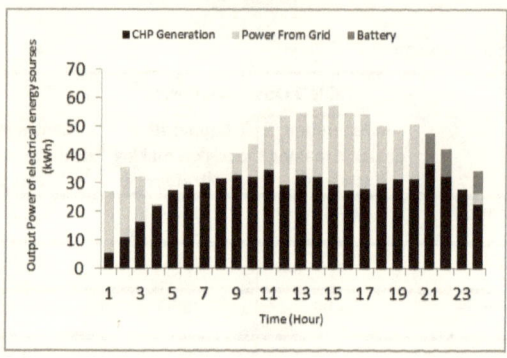

Fig 5 Composition of Electrical Energy Supplies for Hospital in First Scenario

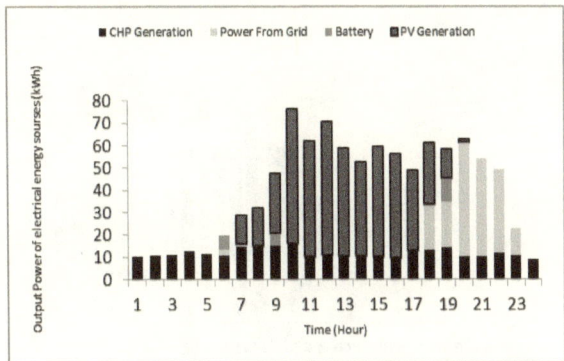

Fig 6 Composition of Electrical Energy Supplies for Commercial Building in First Scenario

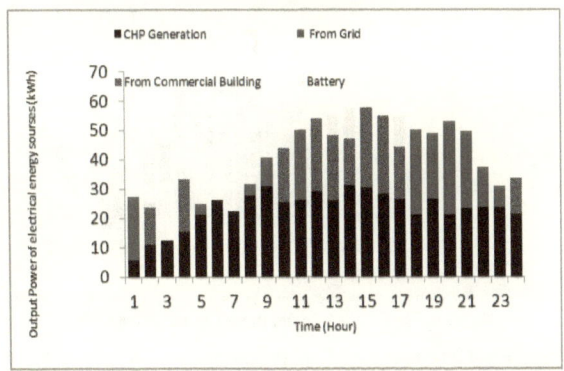

Fig 7 Composition of Electrical Energy Supplies for Hospital in Second Scenario

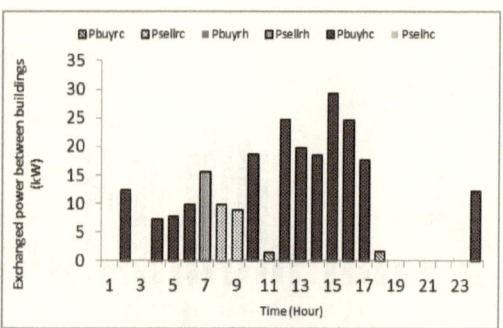

Fig 8Inter Agent Exchanged Power in Second Scenario

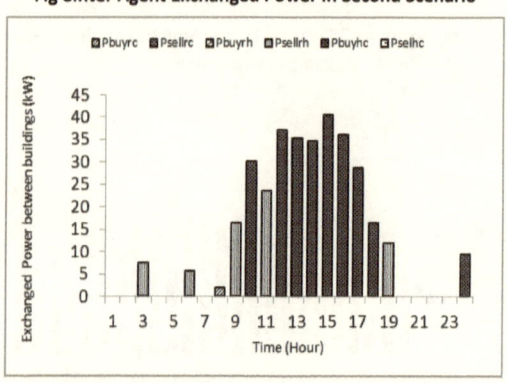

Fig 9Inter Agent Exchanged Power in third Scenario

According to Table 4 in the second scenario, Independent decision making strategy with cooperation among neighbors, by smart controlling the operation cost of each building and therefor the total operation cost would reduce. The DERs productions of hospital building are shown in Fig. 7 and in Fig. 8 the "inter agent exchanged" power has been illustrated.

Table 4 also shows a more reduction in total operation cost in third scenario, Centralized Operation, but besides there is a rise in residential building's cost which is not a good thing. The next section will discuss about different aspects of each operation scenario. Fig. 9 shows the "inter agent exchanged" power in third scenario.

1.6 Discussion

As it can be seen from Table 4the total operation costs are decreasing from first to third scenario. While utilizing centralized rather than independent or cooperative decision making strategies, can bring a cost reduction of 28.14 ($/day) or 15.52 ($/day) respectively, the centralized strategy can jeopardize the privacy of the buildings. The centralized decision making strategy is useful where the buildings have single owner and profit and loss are related to that owner. As it can be seen from Table 4 in third scenario, although the total cost are reduced but, the operation cost of residential building and hospital has risen compared with second scenario. In real conditions the residential and hospital will not participate in such a situation, this centralized operation is useful when the three buildings have one owner. The second scenario, cooperative decision making strategy, is much closer to reality. It has reasonable cost reduction ($15.52 per day equal to 7.4% reduction) and preserves the costumer's privacy and independency in decision making. This data exchange capability can bring cost reduction for buildings. Additionally, this approach can be beneficial in planning new urban areas in order to match the demand of different building types with generation resources. From this point of view determining proper location of buildings with different applications such as residential, commercial, hospital or educational with different electrical load patterns and generation in an area can affect the city planning and design and also the economic size of generating units.

1.7 Conclusion

In this paper optimal resource scheduling for three different buildings has been performed based on communication and cooperation among buildings. The proposed method not only brings up cost reduction in compare with independent planning but also will reduce the need for new resources. Numerical tests showed that it would be significant reductions in total operation cost by implementing cooperative decision making strategy rather than independent operation of buildings. Much of the total operation cost is observed in centralized decision making strategy, but this strategy exposes the privacy of costumers to risk. The proposed idea can be used to form new virtual small smart grid inside real main smart grids and also can be adopted for planning and designing new cities and urban areas.

Reference

1. Basu, A. Bhattacharya, S. Chowdhury, and S. Chowdhury, "Plannedscheduling for economicpower sharing in a CHP-based micro-grid ", IEEE Trans. Power Systems, vol. 27, no. 1, pp. 30 –38, Feb 2012.

2. M. Liserre, T. Sauter, and J. Y. Hung, "Future energy systems, integrating renewable energy sources into the smart power grid through industrial electronics," *IEEE Ind. Electron. Mag.*, vol. 4, no. 1, pp. 18–37, Mar. 2010.

3. R. Calvin, K. Yeager, J. Stuller, 'Power Perfect". McGrawHill,2006

4. R. Lasseter, "Microgrids," inPower Engineering Society Winter Meeting,2002. IEEE, vol. 1, 2002, pp. 305–308 vol.1

5. Mitra, J., and S. Suryanarayanan. "System analytics for smart microgrids." *Power and Energy Society General Meeting, 2010 IEEE.* IEEE, 2010.

6. G. M. Masters, Renewable and Efficient Electric Power Systems. Hoboken, NJ: Wiley, 2004

7. W. Gellings and J. H. Chamberlin, Demand Side Management: Concepts and Methods, 2nd ed. Tulsa, OK: Penn Well Books, 1993.

8. Reducing electricity consumption in houses, Ontario Home Builders' Assoc., May 2006, Energy Conservation Committee Report and Recommendations.

9. Mohsenian-Rad, A., et al. "Autonomous demand-side management based on game-theoretic energy consumption scheduling for the future smart grid." *Smart Grid, IEEE Transactions on* 1.3 (2010): 320-331.

10. X. Guan, Z. Xu, and Q. Jia, "Energy efficient buildings facilitated by microgrid,"IEEE Trans. Smart Grid, vol. 1, no. 3, pp. 243–252, Dec. 2010.

11. Bagherian, Alireza, and SM Moghaddas Tafreshi. "A developed energy management system for a microgrid in the competitive electricity market."*PowerTech, 2009 IEEE Bucharest*. IEEE, 2009.

12. Morais, Hugo, PéterKádár, Pedro Faria, Zita A. Vale, and H. M. Khodr. "Optimal scheduling of a renewable micro-grid in an isolated load area using mixed-integer linear pogramming." *Renewable Energy* 35, no. 1 (2010): 151-156.

13. M. Houwing, R. Negenborn, and B. De Schutter, "Demand response with micro-CHP systems, "Proceedings of the IEEE, vol. 99, no. 1, pp. 200–213, 2011

14. M. Tasdighi, P. JamborSalamati, A. Rahimikian, and H. Ghasemi, "Energy management in a smart residential building," inEnvironment and Electrical Engineering (EEEIC), 2012 11th International Conference on, 2012, pp. 128–133.

15. Tasdighi, Mohammad, Hassan Ghasemi, and AshkanRahimi-Kian. "Residential Microgrid Scheduling Based on Smart Meters Data and Temperature Dependent Thermal Load Modeling." (2013): 1-9.

16. Zhang, Di, Nilay Shah, and Lazaros G. Papageorgiou. "Efficient energy consumption and operation management in a smart building with microgrid." *Energy Conversion and Management* 74 (2013): 209-222.

17. N. Lu, T. Taylor, W. Jiang, J. Correia, L. R. Leung, and P. C. Wong, "The temperature sensitivity of the residential load and commercial building load," presented at the 2009 IEEE Power Energy Soc. Gen. Meet., Calgary, AB, Canada, PESGM2009-000775.

18. H. S. Cho, T. Yamazaki, and M. Hahn, "AERO: extraction of user's activities from electric power consumption data,"IEEE Trans. Consumer Electronics, vol. 56, no. 3, pp. 2011–2018, Sep. 2011
Hajimiragha, C. A. Ca ˜ nizares, M. W. Fowler, and A. Elkami, "Optimal transition to plug-in hybrid electric vehicles in ontario, canada, considering the electricity-grid limitations," IEEE Trans. Industrial Electronics, vol. 57, no. 2, pp. 690–701, Feb. 2010

19. J. Heo, C. Seon Hong, S. Bong Kang, and S. SooJeon, "Design and implementation of control mechanism for standby power reduction," IEEE Trans. Consumer Electronics, vol. 54, no. 1, pp. 179–185, Feb. 2008

20. "Micro-CHP systems for residential applications final report," United Technologies Research Center, 411 silver lane, east hartford, CT 06108, Tech. Rep., Jun. 2006. [Online]. Available: http://nechpi.org/wp-content/uploads/2012/06/MICRO-CHP-UTC.pdf

Chapter 2
Solar PV Technology and Systems

M. Rizwan and Priyanka Chaudhary

2.1. Introduction

Solar energy can be exploited through the solar thermal and solar photovoltaic (SPV) routes for the various applications. SPV technology enables direct conversion of sunlight into electricity through semiconductor devices called solar cells. Solar cells are interconnected and hermetically sealed to constitute a photovoltaic module. The photovoltaic modules are integrated with other components such as storage batteries to constitute SPV systems and power plants.

2.2. Solar Cell

The photovoltaic (PV) cell is the smallest constituent in a photovoltaic system. Photovoltaic Cells are basically made up of a PN junction. When solar irradiance strikes at the cell surface, the photons are absorbed by the atoms of semiconductor; and electrons get free from the negative layer. These free electrons find their way through an external circuit toward the positive layer making flow of electric current from the positive to the negative layer.

With the growing demand of solar power new technologies are being introduced and existing technologies are developing. There are some commonly used solar PV cells are given as following:

❖ Single crystalline or mono crystalline

- ❖ Multi- or poly-crystalline
- ❖ Thin film
- ❖ Amorphous silicon

Single-crystalline or mono crystalline: These are widely available and the most efficient cell materials. They produce the most power per square foot of module. Each cellis made up from a single crystal. To maximize the cell counts in module, the crystals are further divided into rectangular parts.

Polycrystalline cells: It is made from similar silicon material except that instead of being grown into a single crystal, they are melted and poured into a mold. This mold makes square block which can be divided into square wafers resulting in less waste of space and material than round single-crystal wafers.

Thin film panels: Thin film panels are the newest technology in the field of solar cell technology. Thin film materials are Copper indium dieseline, cadmium telluride, and gallium arsenide. They are directly deposited on glass, stainless steel, or other compatible substrate materials. Most of the thin film materials perform slightly better than crystalline modules under low sun light conditions. Thickness of thin film is few micrometers or less.

Amorphous Silicon: This technology is newest in the area of thin film technology. In this technology amorphous silicon vapor is deposited on a couple of micro meter thick amorphous films on stainless steel rolls. This technology utilizes only 1% of the material in comparison with crystalline silicon.

Efficiency of different type of solar cells

Solar Photovoltaic Technology	Solar Cell Type	Materials used	Efficiency (%)
Crystalline Silicon (c-Si) solar cells	Mono Crystalline	Mono Crystalline	14-16
	Poly or Multi crystalline Crystalline	Multi crystalline Crystalline	14-16
Thin film solar cell	Amorphous Silicon (a-Si)	Amorphous Silicon	6-9
	Cadmium Telluride Cd-Te	Cadmium and Tellurium	8-11
	Copper Indium Gallium Selenide (CIGS)	Copper, Indium, Gallium and Selenide	8-11
Multi-junction solar cell	Ga As/Ge/Gallium indium phosphide/gallium arsenide/Germanium	Gallium, Arsenic, Indium, Phosphorus and Germanium	30-35

2.3. Solar Photovoltaic Systems

Solar system design can be design in different manners. But there are two basic design considerations, they are-

1. Standalone
2. Grid connected

1. **Standalone System:** These systems can be utilized to feed power to small loads, like water pumps and street lights and to the huge loads of a house. The main components of a stand-alone system include solar panels, a charge controller and batteries. An inverter

can be added to the design for loads that demands AC power. Output voltage of a panel can be control by an MPPT controller in order to increase the efficiency of the power deliver batteries and load. The components of system vary according to the load requirements and the number of hours for operation during the night. Depending on the load operating during the day the battery may only need to last minutes to hours. For the systems that have loads operating during night, batteries are selected according to the number of hours of operation. Continuous operation of system requires the knowledge of dependability of the load to calculate the amount of reserve energy the system must have to provide. The stand-alone system has number of advantages like independency from the utility grid, replacement of petroleum-fueled generators and provides cost effective solution to remote areas. The main disadvantages are the cost and replacement of equipment and the loss of power during periods of poor solar irradiance.

2. **Grid Connected:** Grid connected systems are mainly composed of a number of PV arrays, which convert the sunlight to DC power and a power electronics unit that changes the DC power to AC power. The produced AC power is fed to the grid and utilized by the local loads. Storage devices can be used to improve the availability of the power generated by the SPV systems. Grid integrated systems have the provisions for the customer to sell back the produced power at cost to the utility grid.

Grid connected photovoltaic systems are gaining more and more attention during the past decade because it has the advantage of more effective utilization of generated power. However, there are number of technical requirements which need to be satisfied from both sides. Some conditions which must be satisfied are:

1. Frequency of PV plant should be same as utility
2. Voltage level should be same for both sides
3. For synchronization it is utmost important that the phase sequence of PV system should be same with utility grid.

The system mainly consists of PV array, DC-DC converter with MPPT control and a single phase H-bridge inverter along with control algorithm. The block diagram of the proposed system is presented in Fig.2.1. A DC-DC boost converter is used to step up the voltage level of PV array. MPPT control includes P & O and a modified P & O algorithm which are used to track the MPP of solar PV system.

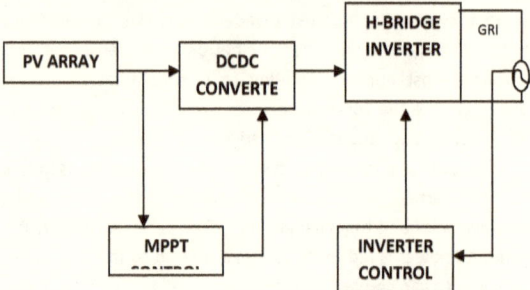

Fig. 2.1 Single phase grid connected PV system

2.4. Component Modeling

The Photovoltaic system mainly consists of PV array, DC-DC converter and battery bank along with MPPT controls. The block diagram of the proposed system is presented in Fig.2.2.

31

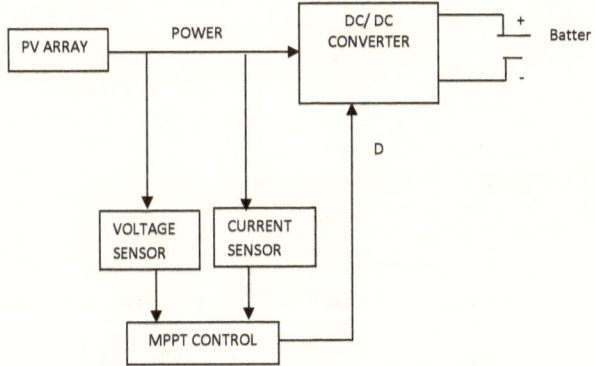

Fig. 2.2 Block diagram standalone of PV system

A DC-DC buck converter is used to step down the voltage level of PV array. Voltage and current sensors are used to sense the current and voltage of PV and different MPPT control includes P & O, incremental conductance and constant voltage are used in this work.

2.5. Pv Array Modeling

PV technology enables direct conversion of sunlight to electricity through semiconductor devices called solar cells. The power produced by solar cell is not enough for power generation applications. To obtain the higher power, solar cells are interconnected and hermetically sealed to constitute a photovoltaic module. Further, series – parallel combination of PV modules constitute a PV array. The equivalent circuit of PV module is presented in Fig. 2.3.

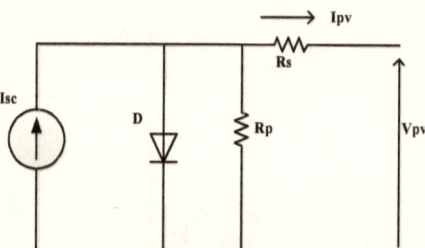

Fig. 2.3 Equivalent circuit model of solar module

From the above circuit diagram, the I-V characteristics can be obtained by

$$I_{sc} - I_D - \frac{V_D}{R_p} - I_{PV} = 0 \qquad (2.1)$$

Thus,

$$I_{PV} = I_{sc} - I_D - \frac{V_D}{R_p} \qquad (2.2)$$

The reverse saturation current I_{rs} is given as

$$I_{rs} = I_{scref} + \left[\exp\left(\frac{qVoc}{NskAT}\right) - 1\right] \qquad (2.3)$$

The module saturation current varies with the cell temperature is given by

$$I_o = I_{rs} \left[\frac{\left(\frac{T}{T_{ref}}\right)^3 e^{qCg}}{Ak} * \left(\frac{1}{T_{ref}} - \frac{1}{T}\right)\right] \qquad (2.4)$$

The basic equation that describes the current output of the PV module of the single-diode model is given in equation (4.5).

$$I_{PV} = I_{sc}N_p - N_s I_o \left[\exp\left\{\frac{q(V_{PV} + I_{PV}Rs)}{N_s AkT}\right\} - 1\right]V_{PV} + \frac{I_{PV}Rs}{Rp} \qquad (2.5)$$

33

Where k is the Boltzmann constant (1.38×10^{-23} J K^{-1}), q is the electronic charge (1.602×10^{-19} C), T is the cell temperature (K); A is the diode ideality factor, R_s the series resistance (Ω) and R_p is the shunt resistance (Ω). N_S is the number of cells connected in series = 72. N_p is the number of cells connected in parallel = 1.

2.6. Dc - Dc Buck Converter

DC – DC converter is used to interface the PV array to dc bus to perform three major functions including step up/step down the PV voltage, regulate the varying dc output voltage of PV array and implement the MPPT of solar array to ensure operation at maximum efficiency. However, there are various topologies of DC-DC converter including buck, boost, push pull, half bridge, full bridge, flyback, buck-boost etc. The choice of topology depends on system requirements and its applications. In this paper, DC-DC buck converter is designed to step down the PV voltage. The circuit diagram for buck converter is presented in Fig. 2.4, and the data of various parameters of buck converter are presented in Table.

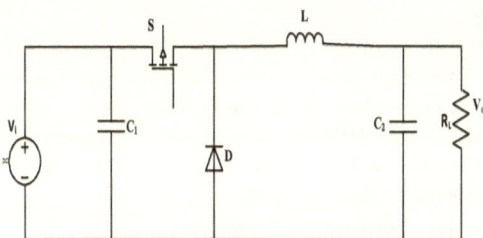

Fig. 2.4DC - DC buck converter

Various parameters of buck converter

S. No.	Parameter	Formula
1.	Input Capacitance (C_1)	$\dfrac{I_{PV}(1-D)}{2\Delta V_{PV}f_{sw}}$
2.	Inductance (L)	$\dfrac{V_{PV}(1-D)}{2\Delta i_L f_{sw}}$
3.	Output Capacitance (C_2)	$\dfrac{\Delta i_L}{8\Delta V f_{sw}}$

2.7. Battery Storage System

The power generated from PV system is fluctuating in nature, therefore there is a need of storage systems. In addition, energy storage systems are useful in both utility and small scale applications. The battery storage systems in grid connected systems helps to achieve the objectives like mitigation of the variability and intermittency of PV power by ensuring the maintenance of constant voltage and frequency meeting the peak electricity demand during low power generation from PV system. There are many battery models are available in literature.

2.8. Maximum Power Point Tracking Techniques

Maximum power point tracking is necessary in order to track the maximum power point (MPP) under varying meteorological parameters. These MPPT techniques are based on the reference voltage or reference current signal of the PV system which is adjusted in order to achieve maximum power point.

(i).Perturb and Observe (P & O) technique

Perturb and observe method is simple method which can be implemented by applying perturbation to the reference voltage or reference current signal of the solar PV system. After the application of perturbation the output power is compared with the previous perturbation cycle power output. If the power increases then the increment in voltage or current remains continuous in same direction. If power decreases then the variation in voltage or current in reverse direction. This method is also known as 'hill climbing method'. A flowchart illustrating this method is shown in Fig. 2.5.

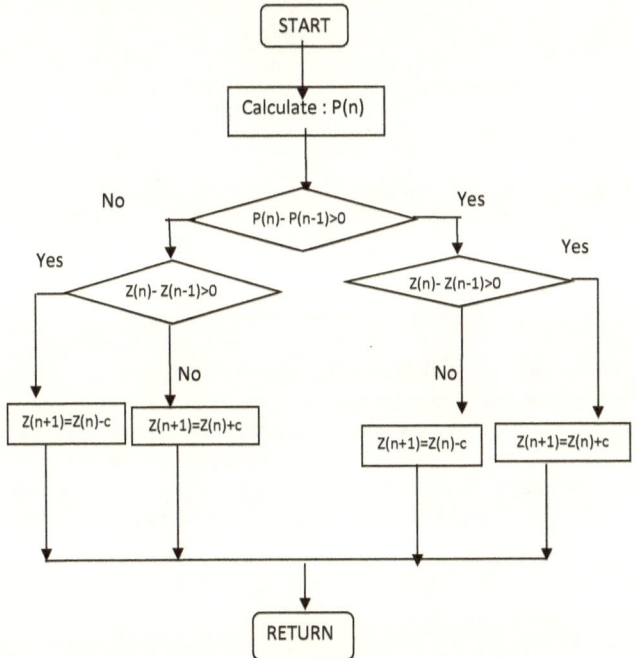

Fig. 2.5 Flowchart for Perturb & Observe MPPT technique

2722727

272727272

(ii). Incremental conductance method

Incremental conductance method is based on the fact that the slope of the PV module curve is zero at maximum power point. This slope will be positive for values of output power smaller then MPP and negative for output power greater then maximum power point.

Maximum output power can be obtained by using the derivative of PV output power with respect to voltage and equating this to zero.

$$\frac{dP}{dV} = I + v\frac{dI}{dV} = 0 \tag{2.6}$$

By using equation (4.6) the following equation can be obtained

$$\frac{dI}{dV} \approx \frac{\Delta I}{\Delta V} = -\frac{I_{MPP}}{V_{MPP}} \tag{2.7}$$

$$\frac{dP}{dV} = 0 \qquad \frac{\Delta I}{\Delta V} = -\frac{I}{V} \qquad at\ MPP \tag{4.8}$$

$$\frac{dP}{dV} > 0 \qquad \frac{\Delta I}{\Delta V} > -\frac{I}{V} \qquad left\ side\ of\ MPP \tag{4.9}$$

$$\frac{dP}{dV} < 0 \qquad \frac{\Delta I}{\Delta V} < -\frac{I}{V} \qquad right\ side\ of\ MPP \tag{4.10}$$

Instantaneous conductance is compared with the incremental conductance in order to track the maximum power point. After achieving the MPP, the operation of PV module is forced to remain at this point unless a change in current occurs as a result of varying meteorological parameters which leads variation in MPP. The flowchart for incremental conductance technique is presented in Fig. 2.6.

37

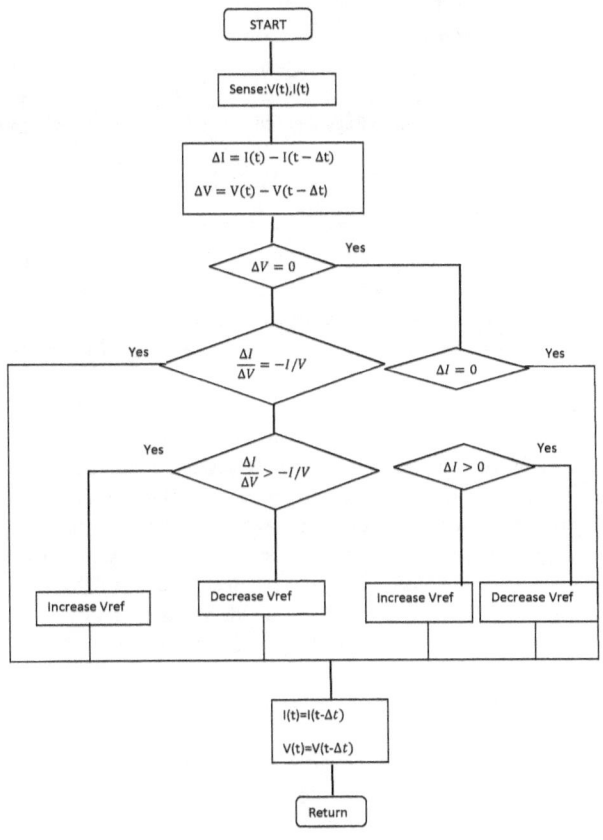

Fig. 2.6 Flowchart for Incremental Conductance MPPT technique

(iii).Constant voltage technique

In this method, operating point of PV module is kept near the maximum power point by regulating the module/array voltage. It should be matched with the fixed reference voltage equal to the V_{MPP} of PV system. The constant voltage algorithm is the simplest MPPT control technique.

2.9. Effect of different parameters on PV system's performance

It is mentioned above that the PV modules are characterized at STC but meteorological parameters like solar irradiance and temperature are not constant under practical conditions. Amount of solar irradiance reaching at earth surface varies greatly because of various meterological and atmoshpheric parameters like water vapour molecules, number of gaseous molecules, aerosoles, cloud, change in position of sun etc. Under such practical conditions, the impact of these parameters is utmost important on PV power output. To incorporate such practical conditions, performance analysis of PV system at varying irradiance and constant temperature, constant irradiance and varying temperature and varying irradiance and temerature is carried out. The effect of solar irradiance and temperature on PV output is presented in Fig. 2.7and Fig. 2.8 respectively. It is clearly mentioned from Fig.2.7 and Fig.2.8that there is a large impact of these parameters on PV output.

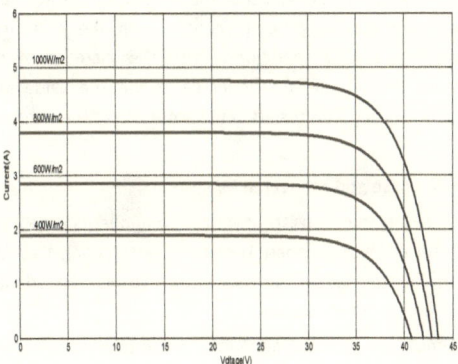

Fig. 2.7 Performance of PV system with varying irradiance

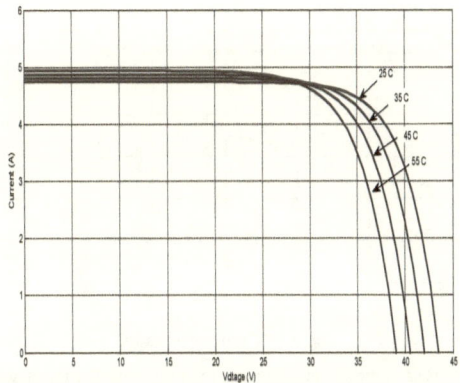

Fig. 2.8 Performance of PV system with varying temperature

Further, it is observed that the PV output current varies with irradiance. The operating point changes with the solar irradiance, temperature and load conditions. With the increase in operating temperature the output current increases but the value of voltage decreases drastically. This results in the reduction of the output power.

2.10. Single phase grid connected PV system

The PV system consists of various components like PV array, DC-DC converter and a single phase H-bridge inverter along it's with control algorithm. The present section shows some results related with single phase grid connected PV system.

Simulation has been done for a single phase grid connected PV system by keeping its parameters at STC. Fig. 2.9 shows the PV output voltage which is maintained at a constant value of 28 V less than the voltage at maximum power point.

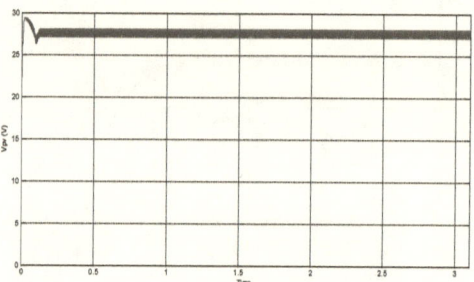

Fig. 2.9PV output voltage

This is utmost important to maintain the DC input voltage of inverter at a constant value, so that the AC output voltage remain constant and the system can be synchronized with the utility grid. Here a reference value of DC link voltage is set at 400 V. In this case the DC link voltage is 380 V approximately and has a constant value throughout the simulation period.

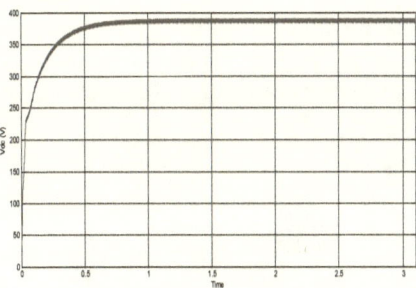

Fig. 2.10DC link voltage (V)

Fig. 2.11 shows the waveform of the grid voltage and Fig. 2.12 grid current (I_{grid}), which is a pure sinusoidal waveform in phase with the grid voltage and at unity power factor. V_{inv} is shown in Fig. 2.13 which is a unipolar square wave voltage. PI control signal is compared with a triangular carrier signal in order to generate gating pulses for the inverter switches.

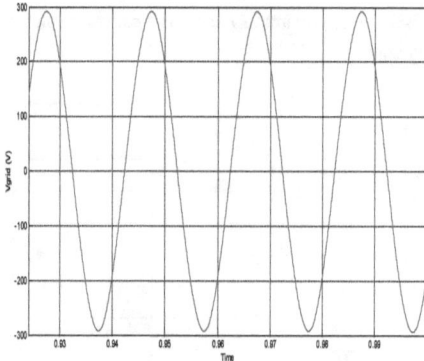

Fig. 2.11Grid voltage (V)

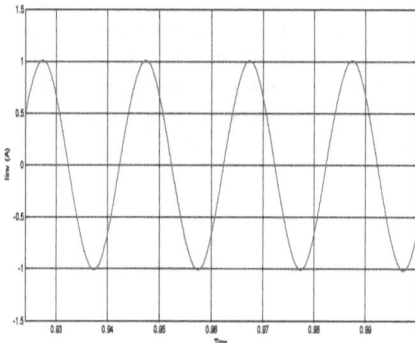

Fig. 2.12Grid current (A)

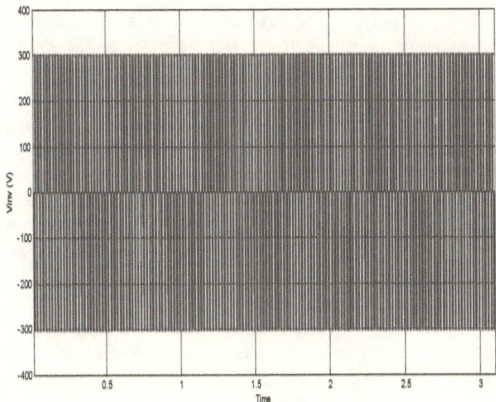

Fig. 2.13Inverter output current (A)

References

1. A R Reisi, Moradi M H and S Jamsab, "Classification and comparison of maximum power point tracking techniques for photovoltaic system: A review" Renewable and Sustainable Energy Reviews 19, 433-443, 2013.
2. Chetan Singh Solanki "Solar Photovoltaics Fundamentals, Technologies and Applications" PHI Learning Private Limited, Second Edition, Reprint 2012.
3. N Pandiarajan, R Ramaprabha and M Ranganath "Application of circuit model for photovoltaic energy conversion system" International Journal of Advanced Engineering Technology, 2(4): 118-127, 2011.
4. N Pandiarajan and M Ranganath "Mathematical modeling of Photovoltaic module with Simulink, IEEE.pp. 258-263, 2011.
5. A Chouder "Modeling and simulation of a grid connected PV systems based on the evaluation of main PV module parameters" Sciverse Science-Direct, 2011.
6. M Rizwan, M Jamil and D P Kothari "Solar energy estimation using REST model for PV-ECS based distributed power generating system", Solar Energy Materials & Solar Cells, Volume 94, pp. 1324-1328, 2010.
7. I H Altas and A M Sharaf "A Photovoltaic Array Simulation Model for Matlab-Simulink GUI Environment" IEEE, Clean Electrical Power, International Conference on Clean Electrical Power (ICCEP '07), Ischia,Italy, June 14-16, 2007.
8. S Jain and V Agarwal, "Comparison of the performance of maximum power point tracking schemes applied to single-stage grid connected photovoltaic systems," IET Electr. Power Appl., vol. 1, no.5, pp. 753–762, Sep. 2007.
9. S B Kjaer, J K Pedersen and FBlaabjerg "A review of single phase grid-connected inverters for photovoltaic modules," IEEE Transactions on Industry Applications, vol. 41, no. 5, pp. 1292-1306,Sep/Oct. 2005.

Chapter 3

Optimal Multi-Objective Reconfiguration in Distribution Systems Using the Novel Fuzzy-Exiwo Method

Hajar Bagheri Tolabi, Mahmoud Reza Shakarami and

Rahil Hosseini

Abstract

This paper presents a new technique for optimal multi-objective reconfiguration of distribution system based on the Expanded Invasive Weed Optimization (exIWO) as a meta-heuristic algorithm. To avoid the convergence problem, the input and output data are normalized in the same range based on fuzzy sets. The purpose of this research includes active power loss reduction, voltage profile improvement and increase of the system load balancing. The proposed method is investigated using IEEE 33-bus test system and a real distribution network i.e. Tai-Power 11.4-kV distribution system. The obtained results showed the proposed fuzzy-ex-IWO method is more accurate in compared to exIWO, and other intelligent search algorithms such as Genetic Algorithm (GA) or Particle Swarm Optimization (PSO) meanwhile it has an efficient convergence property in compared with other intelligent search algorithms.

3.1 Introduction

Electrical power distribution systems have two types of switches: tie and sectionalizing, which determine the configuration of distribution network [1]. By changing the switches statue and transition of sections between feeders during operation, the construction of network distribution will change. This change is known as reconfiguration and is performed from time to time. The main objectives of reconfiguration of any distribution systems include: power loss reduction, improving voltage profile, increase the network reliability and relieve overload in the network [2]. Reconfiguration of distribution network for loss reduction was first proposed by Merlin and Back [3]. They have used a branch and bound optimization method to determine the configuration that has the minimum total loss. In this method, all switches are first closed to establish a meshed configuration.

The switches are then opened successively to achieve the radial configuration. Goswami and Basu [4] presented a heuristic algorithm for reconfiguration which uses a power flow program. Gomes et al. [5] offered an algorithm based on mitigation of the system power loss for the great distribution systems. In 2005, a path to node based reconfiguration modeling has been proposed by Ramoset and Exposito [6]. Zhou et al. combined the heuristic rules and fuzzy logics to solve the reconfiguration problem [7]. The operation aspect of optimal reconfiguration was presented by Borozan and Rajakovic [8]. Nara et al. have solved distribution reconfiguration problem for minimum loss using Genetic Algorithm (GA) [9]. A fuzzy multi-objective reconfiguration approach was offered by Das [10]. The main objectives of this study are increasing the load balancing of the feeders and reducing the active power losses. Rao et al. used Harmony Search Algorithm (HSA) to solve the network reconfiguration problem. The aim of this reconfiguration is to achieve an optimal switching combination simultaneously minimization of the network's active power losses [11].

In this paper, the novel Expanded Invasive Weed Optimization (exIWO) approach is used for the distribution network reconfiguration in order to power losses reduction, improving voltage profile and increasing the load balancing of the feeders. A fuzzy logic technique is used to achieve a compromise between the different objectives. The remainder of this paper is organized as follows: Section 2 presents the problem formulation. Section 3 introduces the exIWO approach. Section 4 explains the optimization process using the proposed fuzzy-exIWO approach. Section 5 presents the simulation and results and section 6 outlines conclusions.

3.2 Problem formulation

One of the main problems in the distribution system is how to find an optimal configuration of the system. For this, an objective function, Fit(x), is define as a constrained optimization problem. This multi-objective function includes three goals: 1) decrease in the network loss, 2) increase the load balancing, and 3) voltage improvement which is formulated as follows:

$$MinFit(X) = \min \{P_{loss}, LBI, VPI\} \qquad (1)$$

with the following constraints:

$1 : V_{k\min} \le V_k^l \le V_{k\max}$

$2 : \left| I_{k,k+1}^l \right| \le \left| I_{k,k+1\max} \right|$

$3 :$ Radial structure of network should be maintained

$4 :$ All available nodes of considered distribution
system should be fed.

where,

V'_k : Voltage at bus k after reconfiguration.

$V_{k\max}$: Maximum bus voltage.

$V_{k\min}$: Minimum bus voltage.

$I'_{k,k+1}$: Current in line section between buses k
and k + 1 after reconfiguration.

$I_{k,k+1\max}$: Maximum current limit of line section
between buses k and k + 1.

In equation 1, the P_{loss} represents real power losses which is defined as follows:

$$P_{loss} = \sum_{k=1}^{n_f} R_k \frac{P_k^2 + Q_k^2}{V_k^2} \qquad (2)$$

where,

V_k : Voltage amplitude at bus k.

P_k : Real power flowing out of bus k,

Q_k : The reactive power flowing out of bus k,

nf : Total number of lines sections.

The second criteria in equation 1, is the Load Balancing Index (LBI) of the lines in the feeder, as follows:

$$LBI = \sum_{F_j} (\frac{I_{Fj}}{I_{Favg}})^2 \qquad (3)$$

where, I_{Fj} is the current passing through line j and I_{Favg} is defined by Eq. (4):

$$I_{Favg} = \frac{1}{n_f} \sum_{j=1}^{n_f} I_{Fj} \qquad (4)$$

49

The last criteria in the equation 1, represents the improvement of the Voltage Profile Index (VPI) is defined as follows:

$$VPI = \sum_{k \in LB} |V_k - V_{ref,k}| \qquad (5)$$

where *LB* shows collection of the load buses, and $V_{ref,k}$ is the nominal voltage at load bus *k*.

The profile of voltages in the feeder buses improves if the VPI index degrades.

3.2.2 Homogenization based on fuzzy system

The criteria in the multi-objective function are in various ranges. For this, a fuzzy system is employed for comparison of the objectives during solving the optimization reconfiguration. In the fuzzy system a membership function (μ) is designed for each objective fuzzy sets, which represents its effectiveness. The membership of each objective is a number in interval [0 1].

The membership function of an objective function is defined as follows:

$$\mu_{fi}(X) = \begin{cases} 1, & f_i(X) \le f_i^{min} \\ \dfrac{f_i^{max} - f_i(X)}{f_i^{max} - f_i^{min}}, & f_i^{main} < f_i(X) < f_i^{max} \\ 0, & f_i^{max} \le f_i(X) \end{cases} \qquad (6)$$

where f_i is the objective function index for (*i=1, 2, 3*), f_i^{min}, represents the best case and f_i^{max} represents the worst solutions in the process of the single-objective optimization.

Furthermore, a fuzzy multi-objective function, *F(X)*, is designed to integrate the fuzzy membership functions associated to each objective through a weighting factor as follows:

$$F(X) = \sum_{i=1}^{3} w_i . \mu_{fij}(X) \qquad\qquad (7)$$

where w_i is the weight associated to the ith objective function. Theses weights are considered equally, $w_1 = w_2 = w_3 = 0.33$, due to the equal importance of three objective functions. In which the three objectives are considered to have equal importance.

3.3. Expanded Invasive Weed Optimization

The metaheuristic Expanded Invasive Weed Optimization (exIWO) algorithm approach explores the search space by transforming a complete solution of the considered problem into another solution.

The exIWO strategy is defined as follows [12]:

$$x[t + 1] = s(v(x[t])) \qquad\qquad (8)$$

where the $x[t+1]$ is the population in the time instant $t + 1$ and $x[t]$ is previous population and "v" and "s" are operators of variation and selection, respectively. The first population, $x[0]$, needs to be initialized, as well.

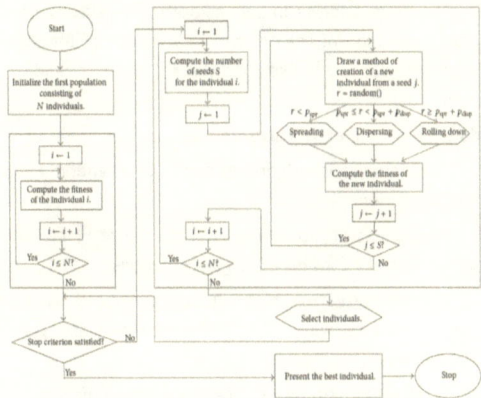

Fig. 1. The flowchart of the ex-IWO algorithm

The simplified pseudo code of the exIWO algorithm is summarized as the following steps:

Step 1: Generate the initial population.

Step 2: For each individual: Calculate the fitness function value.

Step 3: While a termination condition is not met:

Step 3.1: For each individual:

-Calculate the number of seeds.

Step 3.2: For each seed:

- Draw the dissemination method (dispersing, spreading and rolling down).
- Generate a new individual according to the selected method.
- Calculate the fitness function value.

Step 4: Choose individuals to create a new population.

Step 5: Introduce the best individual.

The terminologies in this algorithm is in agreement with the "natural" inspiration of the original IWO version of the algorithm, i.e., the terminologies used like "individual," "plant," and "weed" are synonyms.

The flowchart of the exIWO algorithm is illustrated in Fig.1.

This optimization algorithm begins by a random initialization of the first population.

Protoplasts of individuals with chance of being refined in the next generation are considered as good solutions.

The quality of the best solution generated using the exIWO algorithm cannot be worse than the best protoplasts generated in a "controlled" way.

Termination criteria is considered as the maximum number of populations or the execution time.

Similar to the IWO, the reproduction number of seeds Sind by an individual is related to its fitness Find. i.e., the greater the degree of individual's adaptation, the greater its reproduction ability as follows:

$$S_{ind} = S_{min} + [(F_{ind} - F_{min})\frac{S_{max} - S_{min}}{F_{max} - F_{min}}] \quad (9)$$

Where S_{max}, S_{min} are maximum and minimum admissible number of seeds, respectively, generated by the best member of the population (fitness F_{max}) and by the worst member (fitness F_{min}). The fitness function F in formula (10) can be used for the maximization of the evaluation function F.

For choosing the proper number of seeds S_{ind}, the minimized valuation function is considered as cost K as follows:

$$S_{ind} = S_{min} + [(K_{max} - K_{ind})\frac{S_{max} - S_{min}}{K_{max} - K_{min}}] \quad (10)$$

The hybrid search space exploration strategy randomly applies three methods for selection of the each seed: 1) dispersing, 2) spreading, and 3) rolling down.

The probability parameters of related methods, p_{spr}, p_{spr}, and p_{roll}, are estimated where the following condition is satisfied:

$$p_{spr} + p_{spr} + p_{roll} = 1$$

In this algorithm the numbers are randomly generated using a uniform distribution between 0 and 1.

Random generation of new individuals as seeds includes spread of seeds by disseminating them over the search space.

The locations of new weeds are independent of their parent plants of the search space.

During the optimization process, the distance between the parent and the location of the seed on the ground is the basis for the dispersing and the rolling down process. New individuals are randomly generated and are assigned to the elements of the individual structure (e.g., arguments of n dimensional function). The direction of seeds is identified by considering the distance from seed to parent plant. The individual, as a neighbor, is used in the rolling down procedure

The next population candidates are deterministically selected using three following methods:

1) global, 2) offspring-based, and 3) family-based. The next population of the algorithm includes all μ parent plants and all λ their newly created descendants.

The candidate set of the *global* selection includes all μ parent plants and all their new λ descendants. This evolutionary IWO competitive exclusion mechanism is represented as $(\mu + \lambda)$ [12]. However this offspring-based selection (μ, λ), where $\lambda \geq \mu$, does not consist of only the set of λ descendants and may include non-optimal points in the overall search space [12]. The best individual grown in the current population will be stored for the final optimization result.

In the family-based selection, the idea of the *inver-over* operator origination in [13], where each plant in the initial population is a protoplast of another family (consists of a parent weed and its direct descendants). The evolutionary optimization process chooses the best individual of each family as the member of the next population. This assumption plays an important role in the random spreading strategy to stimulate proper initialization of the first population.

In the all abovementioned selection algorithms, the population cardinality is considered constant in all iterations.

3.5 Proposed fuzzy-exIWO method

This section represents application of proposed fuzzy-exIWO method for multi-objective reconfiguration optimization problem. The specification of the locations of the network tie switches indicates a feeder topology. Therefore, without violating the constraints, the first solution is formed as follows:

$$S^1 = [proposedtieswitches^1]$$

After updating exIWO parameters, second, third, and ..., ith solutions are produced according to the new proposed tie switches as follows:

$$S^i = \{proposedtieswitches^i\}$$

For each solution i, power flow program is carried out. Then fuzzy membership functions of the objectives are computed and compared to the previous solution in order to choose a better solution. This procedure repeats until a stop criterion is reached.

The proposed method is summarized in the following steps:

Step 1) Read data from distribution system and initialize the parameters of the exIWO algorithm.

Step 2) Create a S solution considering the constraints of the problem.

Step 3) Run the power flow program using the method presented in [14], calculate three criteria of the multi-objective function P_{loss}, LBI, VPI. Evaluate the membership function value for each objective i.e. $\mu_{P_{loss}}, \mu_{LBI}$, and μ_{VPI}. Compute the membership function value of the overall fuzzified multi-objective function. Store the results.

Step 4) Update the parameters of the exIWO algorithm to create a new solution using the Step 3 routine. If the fuzzified objective function value of the new solution is better than previous one, replace the S vector using new solution.

Step 5) If Itercount < Itermax, then Itercount = Itercount +1 and go to Step 4.

Step 6) Calculate and print the defuzzified solution and stop.

3.6. Simulation results

Based on the proposed methodology, an analytical software tool has been developed in MATLAB environment.

In the network simulation, two scenarios are considered to analyze the superiority of the proposed method as follow:

Scenario I: the base system without reconfiguration;

Scenario II: the base system with multi-objective reconfiguration based on fuzzy-exIWO method;

3.6.1 Test system 1

This test system consists of a 33-bus distribution system with a total load of 3.7 MW and 2.3 MVAr, 5 tie switches and 32 sectionalizing switches. The details of this testing system are described in [2]. For computing the power flow, S_{base} =100 MVA and V_{base} = 12.66 kV are considered. Figure 2 represents single line diagram of the 33 bus testing system.

The results of applying the proposed method on the test system 1 are presented in Table I. It is observed from this Table that base case power loss in the system is 202.5 kW which is reduced to 129.64 kW by applying the reconfiguration based on the proposed fuzzy-exIWO method. The VPI index which is 1.7 for the base case is decreased to 1.2 and LBI index which is 67.71 for the base case is calculated to 42.75 by applying the reconfiguration based on the proposed fuzzy-exIWO approach. Also, as can be seen in the Table I, the worst voltage value is 0.91p.u for the base case, while it is increased to 0.97 for the scenario II. As can be seen from Table I for the first test system, by applying the reconfiguration using the proposed fuzzy-exIWO approach, the active power loss is improved about 35.98%, as compared with the base case. The voltage profile, load balancing, and worst voltage indexes are improved about 29.41%, 36.86% and 6.59% respectively as compared with the base case.

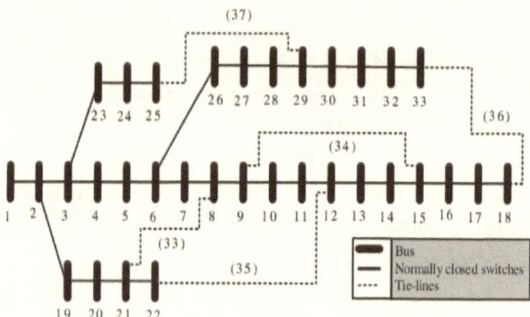

Fig.2. Single line diagram of the 33 bus test system

3.6.2. Test system 2

The second test system employed for the proposed system evaluation is the test system used in the Taiwan power company. Figure 3 represents this test system. The 11.4-kV testing system consists of 83 sectionalizing switches and 13 tie switches. The total load of the system is considered as balanced and constant, which are 28.35 kW and 20.7 kVAr, respectively. More information is provided in [15].

The power flow calculation is performed based on S_{base}=100 MVA and V_{base}= 11.4 kv. The single line diagram of this test system is presented in Table I.

By applying the fuzzy-exIWO method on the test system 2, it is observed that power loss in the base system is 531 kW which is decreased to 356.28 kW through reconfiguration based on the proposed fuzzy-exIWO method. VPI index which is 2.5 for the base case has been reduced to 1.89. LBI index is obtained 140.4 for the scenario I while it is equal to 101.93 for the scenario II. Also the worst voltage value is 0.92 p.u for the base case while it is evaluated 0.96 p.u for scenario II.

As can be seen from Table I for the second test system, by applying the reconfiguration using the proposed fuzzy-exIWO approach, the active power loss is improved about 32.9%, as compared with the base case. The voltage profile, load balancing, and worst voltage indexes are improved about 24.4%, 27.40% and 4.34% respectively as compared with the base case.

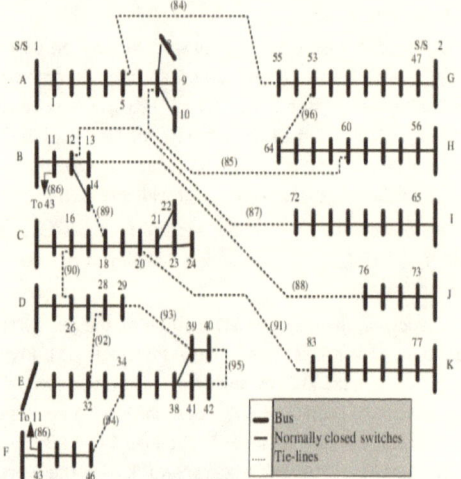

Fig.3. Single line diagram of the Taiwan power company system

Table I. Obtained results for the two test systems

Test system No.	Scenario	Tie switches	Ploss (KW)	Loss reduction (%)	VPI	VPI improvement (%)	LBI	LBI improvement (%)	Worst voltage (p.u.) @ bus	Increment in worst voltage (%)
1	Scenario I	33,34, 35,36,37	202.5	-	1.7	-	67.71	-	0.91 @ 18	-
	Scenario II	33, 9, 34, 28, 36	129.64	35.98	1.2	29.41	42.75	36.86	0.97 @ 32	6.59
2	Scenario I	84, 85, 86, 87, 88, 89, 90, 91, 92, 93, 94, 95, 96	531	-	2.5	-	140.4	-	0.92@10	-
	Scenario II	7, 13, 34, 39, 42, 55, 72, 86, 89, 90, 91, 92, 96	356.28	32.9	1.89	24.4	101.93	27.40	0.96@10	4.34

3.6.3. Comparison of the simulation results by other meta-heuristic methods

Scenario II is simulated based on exIWO, Genetic Algorithm (GA) [13] and Particle Swarm Optimization (PSO) [16] methods to compare with the results obtained by fuzzy-exIWO (proposed method) at nominal load.

As can be seen from Table II, the obtained results based on fuzzy-exIWO is better in comparison with exIWO, GA and PSO in all terms of power loss reduction, voltage profile improvement and increasing the load balancing. The measured P_{loss} based on scenario I, as compared with scenario II based on different intelligent methods i.e. fuzzy-exIWO, exIWO, GA and PSO are presented in figure 4 for the both test systems. As can be seen from this figure, the fuzzy-exIWO method has a lower power loss in compared with other intelligent algorithms. For the other items such as LBI and VPI, as seen in the Table III, better results has been delivered by the proposed method in compared with exIWO (without optimization by fuzzy system), GA and PSO techniques.

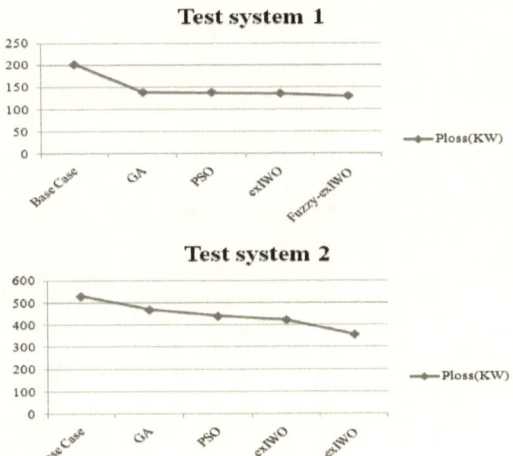

Fig.4. the measured P_{loss} for base case and scenario II based on different methods

Table II. Comparison of the simulation results

	Methods	Case	Scenario II
	GA	Tie-switches	33,9,34,28,36
		P_{loss}	138.49
		VPI	1.597
		LBI	51.98
	PSO	Tie-switches	31,7,9, 14,37
		P_{loss}	138.12
Test system 1		VPI	1.496
		LBI	51.87
	exIWO	Tie-switches	9, 14, 27, 33, 36

		P_{loss}	135.37
		VPI	1.43
		LBI	51.79
	Fuzzy-exIWO	Tie-switches	33, 9, 34, 28, 36
		P_{loss}	129.64
		VPI	1.2
		LBI	42.75
Test system 2	Methods	Case	Scenario II
	GA	Tie-switches	7, 13, 34, 39, 42, 55, 72, 86, 89, 90, 91, 92, 96
		P_{loss}	470.09
		VPI	2.24
		LBI	137.36
	PSO	Tie-switches	7, 13, 34, 39, 41, 61, 84, 86, 87, 89, 90, 91, 92
		P_{loss}	440.7
		VPI	2.09
		LBI	135.16
	exIWO	Tie-switches	7, 13, 34, 39, 41, 61, 84, 86, 87, 89, 90, 91, 92
		P_{loss}	423.62
		VPI	2.08
		LBI	139.57
	Fuzzy-exIWO	Tie-switches	7, 13, 34, 39, 42, 55, 72, 86, 89, 90, 91, 92, 96
		P_{loss}	356.28
		VPI	1.89
		LBI	101.93

3.7. Conclusion

In this paper, a new method has been proposed based on the combination of fuzzy system and expanded invasive weed optimization (fuzzy-exIWO) for the multi-objective reconfiguration in distribution systems. The main objectives include the power loss reduction, voltage profile improvement, and increasing the load balancing of distribution system. The proposed method is validated using the 33-bus test system and a real Tai-Power 11.4-kV distribution system. By applying the proposed method it has been seen that the active power loss is improved about 35.98%, as compared with the base case. The voltage profile, load balancing, and worst voltage indexes are improved about 29.41%, 36.86% and 6.59% respectively as compared with the base case for the 33-bus test system. In the case of Tai-Power distribution test system, the loss is reduced about 32.9%, voltage profile, load balancing, and worst voltage are improved about 24.4%, 27.40% and 4.34%respectively as compared with the base system.

The obtained results are also compared with exIWO, GA and PSO at nominal load. The computational results showed that performance of the proposed fuzzy-exIWO is better than exIWO (without optimization by fuzzy system), GA and PSO methods in all terms of power loss reduction, voltage profile improvement and equalizing the feeder load balancing.

References

1. H. Bagheri Tolabi, M. Ali, M. Rizwan, Simultaneous Reconfiguration, Optimal Placement of DSTATCOM, and Photovoltaic Array in a Distribution System Based on Fuzzy-ACO Approach, IEEE Transactions on Sustainable Energy, vol. 6, no. 1, 2015, pp. 210-218.

2. Hajar Bagheri Tolabi, Mohd Hasan Ali, Shahrin Bin Md Ayob, M. Rizwan, Novel hybrid fuzzy-Bees algorithm for optimal feeder multi-objective reconfiguration by considering multiple-distributed generation, Energy, vol. 71, 2014, pp. 507-515

3. A. Merlin and H. Back. Search for a minimal-loss operating spanning tree configuration in an urban power distribution system. in Proc. 5th Power System Computation Conf., Cambridge, pp.1–18,1975,U.K.

4. S. K. Goswami and S. K. Basu. A new algorithm for the reconfiguration of distribution feeders for loss   minimization. IEEE Trans. Power Del. vol.7, n.3, 1992,pp.1484 –1491.

5. F. V. Gomes., S. Carneiro., and J. L. R Pereira., Garcia Mpvpan, and RamosAraujo, L. A new heuristic reconfiguration algorithm for large distribution systems. IEEE Transactions on Power Systems, vol. 20 n.3, 2005, pp.1373–1378.

6. E. R. Ramos and A. G. Exposito. Path-based distribution network modeling: Application to reconfiguration for loss reduction. IEEE Trans. Power Syst. vol.20 n.2, 2005, pp. 556 –564.

7. Q. Zhou, D. Shirmohammadi and W. H. E. Liu. Distribution Feeder Reconfiguration for Service Restoration and Load Balancing. IEEE Trans. Power Syst. vol.12, n.2,1997,pp. 724–729.

8. V. Borozan and N. Rajakovic. Application Assessments of Distribution Network Minimum Loss Reconfiguration. IEEE Transaction on Power Delivery. vol.12 n.4, 1997, pp.1786–1792.

9. K. Nara, A. Shiose, M. Kitagawa, and T. Ishihara. Implementation Of Genetic Algorithm for Distribution System Loss Minimum Reconfiguration. IEEE Transaction on Power Delivery. vol. 7 n.3, 1992,pp.1044–1051.

10. D. Das. A Fuzzy Multi-Objective Approach for Network Reconfiguration of Distribution Systems. IEEE Transaction on Power Delivery. vol. 21, n.1, 2006,pp.202–209.

11. R.S. Rao, S.V.L. Narasimham, M.R. Raju, and A.S. Rao. Optimal Network Reconfiguration of Large-Scale Distribution System Using Harmony Search Algorithm. IEEE Transactions on Power Systems. vol. 26 n.3, 2011, pp.1080-1088.

12. Z. Michalewicz and D. B. Fogel, How to Solve It: Modern Heuristics, Springer, 2004.

13. G. Tao and Z. Michalewicz, "Inver-over operator for the TSP," in Parallel Problem Solving from Nature—PPSN V, vol. 1498 of Lecture Notes In Computer Science, pp. 803–812, Springer, 1998.

14. S. Ghosh, and K. S. Sherpa, "An efficient method for load flow solution of radial distribution networks," *Int. J. Elect. Power Energy Syst. Eng.* ,vol. 1, no. 2, pp. 108–115, 2008.

15. L.W. Oliveira, S. Carneiro, E.J. Oliveira, J.L.R. Pereira, I.C. Silva, J.S. Costa, "Optimal reconfiguration and capacitor allocation in radial distribution systems for energy losses minimization", Electrical Power and Energy Systems, Vol. 32, pp. 840–848, 2010.

16. J. Olamaei, T. Niknam, and G. Gharehpetian, Application of particle swarm optimization for distribution feeder reconfiguration considering distributed generators, Applied Mathematics and Computation, Vol. 201, No.1–2, 15 July 2008, pp. 575–586.

Chapter 4
Impedance-Source Inverter Application in Differential Evolution Maximum Power Point Tracking

Mohammad Faridun Naim Tajuddin, Shahrin Md Ayob and

Zainal Salam

4.1 Introduction

The impedance-source inverter (ZSI) can still be regarded as new inverter topology as compared with voltage-source and current source inverters. The ZSI was firstly introduced in year 2003 by Fang Zheng. As its name imposed, the ZSI use the combination of inductor and capacitor circuits as the power source. The use of the impedance made the output voltage to be possibly higher or lower by controlling the switching time and the 'shoot-through' of the power semiconductors. The inverter realize the inversion together with boost or buck function in single processing stage [1]. Hence producing, less components, higher reliability and higher efficiency type of power converter.

The employment of ZSI has been vastly reported in power conditioner applications i.e. electrical drives, active power filter and etc [2-5]. In photovoltaic maximum power point tracking (MPPT), the ZSI has been used with conventional MPPT algorithms such as Perturb and Observe (P&O), Incremental Conductance (IncCond), Hill Climbing (HC) and etc. to track the maximum power point.

The conventional MPPT algorithms perform well in uniform shading but will fail to converge to the real value of global MPP point under partial shading condition. In effort to overcome the aforementioned problems, several researchers have utilized the artificial intelligent (AI) techniques such as fuzzy logic control (FLC) [6] and neural network (NN) [7] for the MPPT applications. Both techniques are proven to be very effective in dealing with nonlinear characteristics of solar cell *I-V* curve, but the drawback is that they need an extensive computational and still unable to locate the global peak of MPP.

This chapter will describe in detail the deployments of impedance-source inverter (ZSI) for Differential Evolution (DE) MPPT algorithm. The algorithm can be categorized under evolutionary algorithm (EA) class and it is quite new to MPPT application. The main impetus of using DE is based on studies in other fields that demonstrated that DE converges fast, accurate, robust, simple in implementation and requires only a few control parameters [8].

4.2 The Proposed ZSI with DE MPPT

The ZSI can be used to realize both dc voltage boost/buck and dc-ac inversion in single stage with additional features that cannot be accomplished with the conventional voltage or current source inverters. **Figure 1** shows the full circuit block diagrams that cover the ZSI circuit itself, the photovoltaic modules model and the MPPT where the DE algorithm is developed. An impedance network containing two equal (split) inductors in series and diagonally connected to two equal (split) capacitors, outputs DC voltage (DC link voltage) to a single-phase inverter bridge, which is comprised of four power semiconductors.

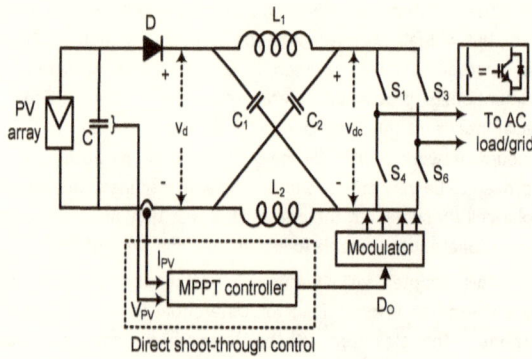

Figure 1: The proposed ZSI with DE MPPT

The MPPT control algorithm provides a shoot-through interval which should be inserted in the switching waveforms of the inverter to output maximum amount of power to the Z-network. At this instant, the voltage across the Z-source capacitor, V_C is equal to the output voltage of the PV array (V_{PV}). A ZSI has three operating modes, namely an active (non shoot-through) mode, a shoot-through mode and a traditional zero mode.

From the symmetry of the Z-source network the following is obtained:

$$V_{C1} = V_{C2} = V_C \; v_{L1} = v_{L2} = v_L \qquad (1)$$

where V_C is voltage across the Z-source capacitor, and v_L is voltage across the inductor.

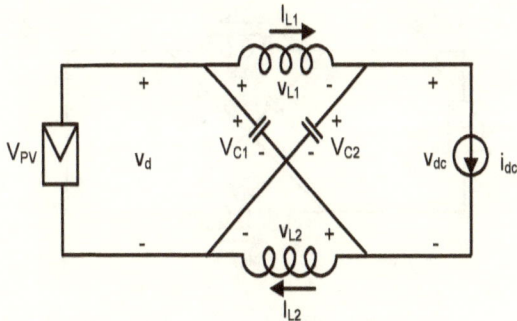

Figure 2: Equivalent circuit of the Z-source inverter viewed from the dc link during non-shoot-through state

Consider that the inverter bridge is in one of the four non-shoot-through switching states, for an interval of T_1. Now the inverter bridge acts as a traditional VSI, thus acting as a current source as shown in **Figure 2**. Due to the symmetrical configuration of the circuit, both of the equal inductors have identical current values. The diode D, shown in the power circuit is forward biased in this case. The voltage across the Z-network in this case can be written as follows:

$$v_L = V_{PV} - V_C \quad v_{dc} = V_{PV}$$
$$\hat{v}_{dc} = V_C - v_L = 2V_C - V_{PV}$$

(2)

Where V_{PV} is the output voltage of the PV array, v_{dc} is the DC link voltage and $\hat{v}_{dc}$ is the peak DC link voltage of the inverter.

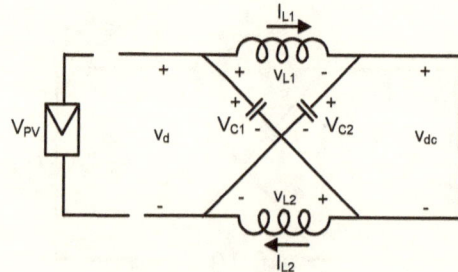

During the non-shoot-through switching state of operation, the inverter bridge can be represented by a current source with zero value (i.e., an open circuit). Therefore, **Figure 3** can represent the equivalent circuit of ZSI from the dc-link point of view when the inverter bridge is in one of the four non-shoot-through switching states.

The inverter bridge is under the shoot-through state for an interval of T_0, during a switching cycle, T. During this mode, the inverter bridge is seen as a short circuit from the DC link point of view. From the equivalent circuit shown in **Figure 3**, the voltage across the impedance elements can be related as:

$$V_L = V_C; V_{PV} = 2V_C; \quad v_{dc} = 0 \tag{3}$$

Figure 3: Equivalent circuit of the Z-source inverter viewed from the dc link when the inverter bridge is in the shoot-through zero state.

In steady state condition, the average inductor voltage over one switching period, T should be zero. Thus from equation (5.2) and (5.3), one has

$$V_L = \overline{v}_L = \frac{T_0 \cdot V_C + T_1 \cdot (V_{PV} - V_C)}{T} = 0 \qquad (4)$$

or

$$\frac{V_C}{V_{PV}} = \frac{T_1}{T_1 - T_0} \qquad (5)$$

where T_1 is a the non-shoot-through period and T_0 is the shoot-through period. Similarly, the average dc-link voltage of the inverter can be written as:

$$V_{DC} = \overline{v}_{dc} = \frac{T_0 \cdot 0 + T_1 \cdot (2V_C - V_{PV})}{T} = \frac{T_1}{T_1 - T_0} V_{PV} = V_C \qquad (6)$$

The peak dc-link voltage across the inverter bridge, expressed in equation (2), can be rewritten as:

$$\hat{v}_{dc} = V_C - v_L = 2V_C - V_{PV} = \frac{T}{T_1 - T_0} V_{PV} = B \cdot V_{PV} \qquad (7)$$

where

$$B = \frac{T}{T_1 - T_0} = \frac{1}{1 - 2(T_0/T)} = \frac{1}{1 - 2D_0} \geq 1 \qquad (8)$$

is the boost factor and D_0 can be referred to as the shoot-through duty ratio and is equal to (T_0/T). On the AC side, the output peak phase voltage from the inverter can be expressed as

$$\hat{v}_{ac} = M \frac{\hat{v}_{dc}}{2} = M \cdot B \cdot \frac{V_{PV}}{2} \qquad (9)$$

where M is the modulation index ($M \leq 1$). By choosing the appropriate buck-boost factor, B_B, the output voltage can be stepped up and down.

$$B_B = M \cdot B = (0 \approx \infty) \tag{10}$$

From equation (1), (5) and (8), the capacitor voltage can be written as

$$V_{C1} = V_{C2} = V_C = \frac{1-(T_0/T)}{1-2(T_0/T)}V_{PV} = \frac{1-D_0}{1-2D_0}V_{PV} \tag{11}$$

4.2.1 Single-Phase ZSI Switching Scheme

It was demonstrated that the shoot-through states of the ZSC can be distributed among the PWM switching pattern of a VSI [1]. It is vital to keep the active states the same as in the VSI PWM switching pattern to avoid voltage waveform distrotion.This section will briefly discussed on the design and development of the shoot-through states that for H-bridge (single phase) ZSI with the used of Simple Boost Control (SBC) method.

Consider the H-bridge ZSI in **Figure 1** whose switching states are given in **Table 1**. The active and null-states, in which the two switches of a phase-leg are switched complementary, are common to both conventional VSI and the H-bridge ZSI. However, the remaining three shoot-through states in which one phase leg (H1 and H2) or two phase-legs (H3) are short-circuited are unique to the H-bridge ZSI [9].

Table 1: Switching States of the H-bridge Z-source inverter ($!S_x$ is complement of S_x)

State (output voltage)	S_1	S_3	S_4	S_6
Active {10} (non-zero)	1	0	0	1
Active {01} (non-zero)	0	1	1	0
Null [28] (zero)	0	1	0	1
Null {11} (zero)	1	0	1	0
Shoot-through H1 (zero)	1	1	S_3	$!S_3$
Shoot-through H2 (zero)	S_1	$!S_1$	1	1
Shoot-through H3 (zero)	1	1	1	1

Figure 4(a) shows the implementation block diagram of SBC method while **Figure 4(b)** shows the modulation pulses of SBC for single phase ZSI with inserted shoot-through periods.

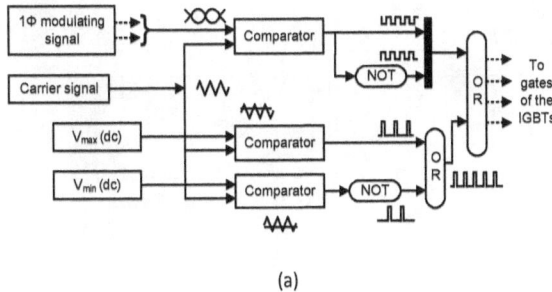

(a)

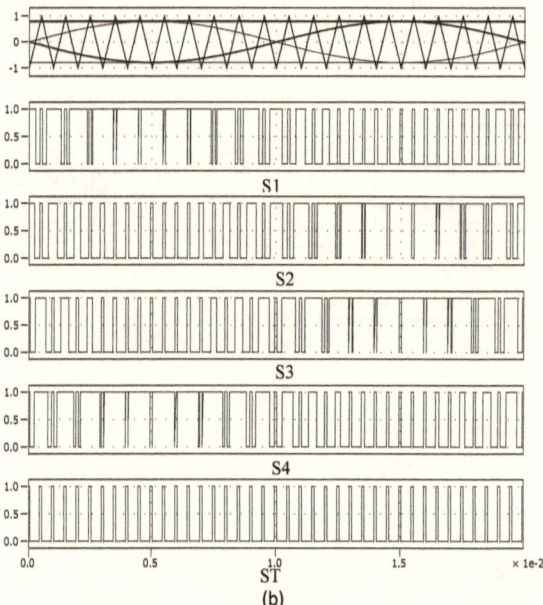

Figure 4: Simple boost control- (a) Implementation logical block diagram and (b) modulated switching waveforms

The two-phase leg of the single-phase ZSI are controlled by carrier based SPWM. The reference signals used are $V_a = Mcos(\omega t)$ for modulating phase-leg $\{S_1, S_4\}$ and $-V_a$ for phase-leg $\{S_3, S_6\}$. Two straight line equal to, or greater than, the peak value of the reference signal, are used to control the shoot-through duty ratio. When the carrier signals is greater than the upper straight line or lower than the bottom straight line, the circuit turns into shoot-through state [10]. ZSI maintain the active states unchanged as in traditional carrier based PWM.

For this simple boosting method, the obtainable shoot-through duty ratio decreases with the increase of the modulation index, M. The maximum shoot-through duty ratio of the simple boosting method is limited to $1 - M$, thus reaching zero at a modulation index of one. In order to produce an output voltage that requires a high voltage gain, a small modulation index has to be used.

4.2.2 The Differential Evolution (DE) MPPT

To ensure the optimal utilization of large PV arrays, maximum power point tracker (MPPT) is employed in conjunction with the power converter. The MPPT control scheme for a ZSI based is shown in **Figure 5.** For the ZSI, the MPPT algorithm generates a shoot-through period (T_0) to boost the Z-source capacitor voltage to the PV array voltage at the MPP. As discussed in the previous section, the shoot-through duty period (T_0) required to boost the capacitor voltage is directly calculated and the shoot-through reference straight lines are generated to produce shoot-through pulses with a simple boost control, as shown in **Figure 4(b).**

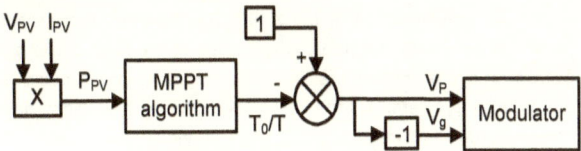

Figure 5: ZSI control block diagram

For DE MPPT design, the key parameters of control are: *NP* - the population size, *CR* - the crossover constant, *F* - the weight applied to random differential (scaling factor). It is worth noting that DE's control variables, *NP*, *F* and *CR*, are not difficult to choose in order to obtain promising results. Storn [11] have come out with several rules in selecting the control parameters. The rules are listed below:

1. The initialized population should be spread as much as possible over the objective function surface.

2. Frequently the crossover probability $CR \in [0,1]$ must be considerably lower than one (e.g. 0.3). If no convergence can be achieved, $CR \in [0.8, 1]$ often helps.

3. For many applications $NP=10 \times D$, where *D* is the number of problem dimension. *F* is usually chosen at [0.5, 1].

4. The higher the population size, *NP*, the lower the weighting factor *F* should choose.

4.3 Simulation Results

The ZSI together with DE MPPT of **Figure 1**is simulated using Matlab /Simulink software. For the impedance network, the following values are choosen ; $L_1 = L_2 = 1$ mH, $C_1 = C_2 = 1000$ μF. The switching frequency is set to 10 kHz. The system is simulated under various irradiance and shading patterns. However, only a prominent result is shown in this chapter due to page limitations.

To show the effectiveness of the ZSI with DE MPPT, the shading pattern as illustrated in **Figure 6** is subjected. It has three power point peaks with one of them is the true global maximum power point. For this type of problem, a good MPPT algorithm with efficient power converter can search and settle at the true global power peak. The result of the tracking is shown in **Figure 7**. The ZSI with DE MPPT is capable to track the global maximum power point with V_{MP}voltage ($17.1 \times 4 = 68.4$ V) and I_{MP}current (3.5 A), respectively. At $t = 1.0$, the operating point is reduced due to lower irradiance. Th new power point is 146.56 Watt. From the result of **Figure 7**, it is shown that at $t = 1.0$, the DE MPPT is capable to track the new power point without a problem. **Figure 8** shows the corresponding output of ZSI.

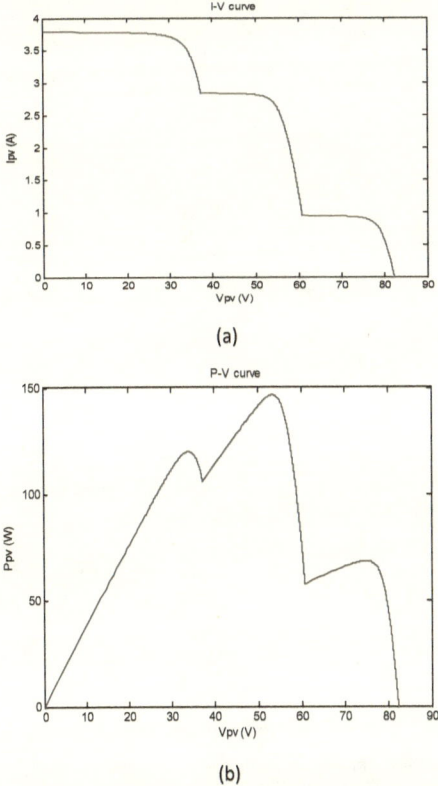

(a)

(b)

Figure 6:*I-V* and *P-V* **curves during partial shading**

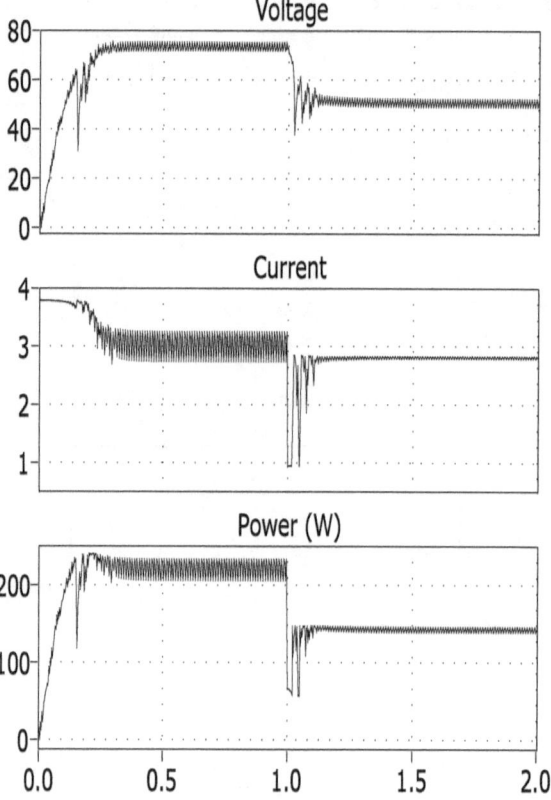

Figure 7: Tracking voltage, current, power of DE MPPT

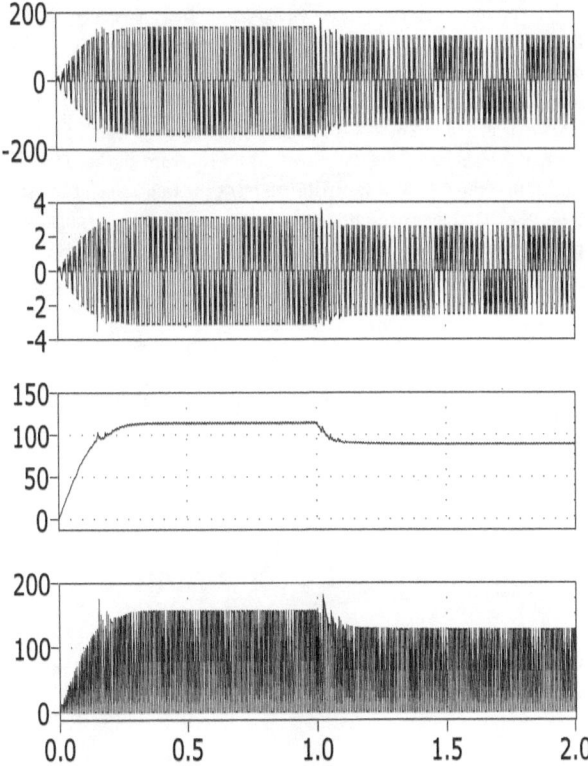

Figure 8: Load voltage V_L, load current I_L, capacitor voltage V_c and inverter voltage V_i of ZSI

4.5 Conclusion

In this chapter, a brief design of impedance-source inverter (ZSI), principle operation and its possible application for MPPT is described. For fast MPP tracking, a Differential Evolution MPPT is employed and integrated with the ZSI. The model is developed in Mat lab/Simulink and the model is subjected with partial shading pattern to verify the viability of the ZSI with DE MPPT. From the result, it was shown that the DE MPPT is capable to track the MPPT and at the same time, the ZSI convert the dc power to an ac power in a single-stage power conversion.

References

1. P. Fang Zheng, "Z-source inverter," Industry Applications, IEEE Transactions on, vol. 39, pp. 504-510, 2003.
2. R. Badin, et al., "Grid Interconnected Z-Source PV System," in Power Electronics Specialists Conference, 2007. PESC 2007. IEEE, 2007, pp. 2328-2333.
3. A. Das, et al., "Residential solar power systems using Z - source inverter," in TENCON 2008 - 2008 IEEE Region 10 Conference, 2008, pp. 1-6.
4. X. Po, et al., "Study of Z-Source Inverter for Grid-Connected PV Systems," in Power Electronics Specialists Conference, 2006. PESC '06. 37th IEEE, 2006, pp. 1-5.
5. L. Yuan, et al., "Controller design for quasi-Z-source inverter in photovoltaic systems," in Energy Conversion Congress and Exposition (ECCE), 2010 IEEE, 2010, pp. 3187-3194.
6. F. Chekired, et al., "Implementation of a MPPT fuzzy controller for photovoltaic systems on FPGA circuit," Energy Procedia, vol. 6, pp. 541-549, 2011.
7. A. K. Rai, et al., "Simulation model of ANN based maximum power point tracking controller for solar PV system," Solar Energy Materials and Solar Cells, vol. 95, pp. 773-778, 2011.
8. M. Miyatake, et al., "Maximum Power Point Tracking of Multiple Photovoltaic Arrays: A PSO Approach," Aerospace and Electronic Systems, IEEE Transactions on, vol. 47, pp. 367-380, 2011.
9. R. Storn, "On the usage of differential evolution for function optimization," in Fuzzy Information Processing Society, 1996. NAFIPS. 1996 Biennial Conference of the North American, 1996, pp. 519-523.
10. L. Poh Chiang, et al., "Pulse-width modulation of Z-source inverters," Power Electronics, IEEE Transactions on, vol. 20, pp. 1346-1355, 2005.
11. R. Storn, "On the Usage of Differential Evolution for Function Optimization," in Proceedings of the Fuzzy Information Processing Society, Berkeley, CA, USA, 1996, pp. 519-523.

Chapter 5
Intelligent Models of Solar Energy

M. Rizwan and Majid Jamil

5.1 Introduction

There are number of mathematical and regression models for the assessment of solar irradiance under cloudless skies available in the literature. These models are not suitable to estimate the solar irradiance during monsoon months or cloudy sky. There are uncertainties in the atmosphere. These uncertainties are due to the existence of dust, moisture, aerosols, clouds, or temperature differences in the lower atmosphere. Of all these factors, clouds can cause the maximum losses in the extraterrestrial solar irradiance reaching the surface at ground level. The atmosphere causes a reduction of the extraterrestrial solar input by about 30% on a very clear day to nearly 100% on a very cloudy day.

As we know that, in India around 50 to 100 days in a year are cloudy, so it is very difficult to predict the accurate results of solar irradiance using mathematical or regression model. Due to uncertainty in weather conditions, a model based on fuzzy logic, artificial neural networks and generalized neural network are presented in this chapter.

5.2 Fuzzy Logic

The concept of Fuzzy Logic was introduced by Professor Lotfi A. Zadeh at the University of California at Berkeley in the 1960's. His goal was to develop a model that could more closely describe the natural language process. This model was intended to be used in situations when deterministic and/or probabilistic models do not provide a realistic description of the phenomenon under study.

This concept provides a natural way of dealing with the problems in which source of in accuracy is the absence of sharply defined criteria rather than the presence of random variables. In this approach, there is no model parameter but all the uncertainties and complications are represented in the form of IF-THEN statements.

5.2.1 Fuzzy Sets

The fuzzy sets and fuzzy operators are the subjects and verbs of fuzzy logic. But in order to say anything useful, we need to make complete sentences. The condition statements, IF-THEN rules, are things that make fuzzy logic useful. The fuzzy logic IF-THEN statements are used to characterize the state of a system and truth value of the proposition is a measure for how well the description matches the state of the system. The fuzzy set can be defined as follows:

Let X, be a universal set. The characteristic function μ_A of a subset of X takes its values in the two element set {0, 1} and is such that $\mu_A(x) = 1$, if x ⊠ A and zero otherwise.

A fuzzy set A has a characteristic function taking its values in the interval {0, 1}. μ_A is the also called a membership function and $\mu_A(x)$ is the grade of membership of x ⊠ X in A. In fuzzy set, the transition between membership and non-membership is gradual rather than abrupt. The union and intersection of two fuzzy subsets A and B of X having membership function μ_A and μ_B respectively is defined as

Union: $\mu_{A \cup B}(x) = \max [\mu_A(x), \mu_B(x)]$ (5.1)

Intersection: $\mu_{A \cap B}(x) = \min [\mu_A(x), \mu_B(x)]$ (5.2)

Fuzzy logic describes the vague concepts such as fast runner, hot weather, weekend days etc. It is convinient way to map an input to an output space. The concept of fuzzy provides a natural way dealing with the problems in which source of impression is the absence of sharply defined criterion rather than the presence of random variables. Prof. Zadeh also introduced linguistic as variables whose values are sentences in natural or artificial language.

5.2.2 Fuzzy Inference Systems

Fuzzy inference systems have been successfully applied in the various fields of engineering such as automatic control, power system analysis, circuit theory, electrical machines and drives etc. It is actual process of mapping from a given input to an output using fuzzy logic. In the fuzzy inference method sets of corresponding input and output measurements are provided to fuzzy system and it learns how to transform a set of inputs to corresponding set of outputs through fuzzy associative map or memory.

5.2.3 Fuzzy Membership Function

The literature is rich enough with references concerning the ways to assign membership values or functions to fuzzy variables. Among them ways are intuition, inference, rank ordering, angular fuzzy sets, neural networks, genetic algorithm, generalized neuron etc. The commonly used approach in fuzzy system is intuitive approach because it is derived from the capacity of humans to develop membership functions through their own innate intelligence and understanding. Intuition involves contextual and semantic knowledge about an issue; it can also involve linguistic truth values about this knowledge. Fuzzy membership is a curve that defined how each point in the input space is mapped to membership value or degree of membership between 0 and 1. Fuzzy membership functions may take many forms, but in practical applications simple linear functions are preferable. In particular, triangular functions with equal base width are the simplest possible choice.

5.2.4 Fuzzy Random Variable

a. Random variable

If an experiment is performed in which one or more variables behave randomly, then the outcome of the experiment will be random. In such a situation, it is not possible to specify in complete detail the outcome of the experiment in advance. Thus the theory of probability can be used to predict the outcome of an experiment affected by random variation in the variables.

Moreover, the outcome of such an experiment is clearly defined; that is, the event corresponding to the experiment will either occur or will not occur at all.

Let Ω be the set of all outcomes of a random experiment. The set Ω is also known as probability space. Then each element ω in Ω specifies exactly what happened; that is, ω is an outcome. A random variable Y is a function from Ω into the set R of real numbers; that is, $Y: \Omega R \rightarrow$. If Y is a random variable, each outcome will specify the value of Y. In other words, associated with each $\omega \in \Omega$, there is a number $Y(\omega)$.

b. Fuzzy variable

Fuzziness does not harbor the concept of clearly defined success (or failure) of outcome of an experiment. Rather it proposes to model the behavior of the outcome of an experiment by allowing the concept of partial success (or partial occurrence).

Let U be a universal set. A fuzzy subset F of U can be defined as a function:

$f_F: U \rightarrow [0, 1]$. The membership function f_F of F takes its value in the interval $[0, 1]$. Let y be an element in U; then $f_F(y)$ is the degree (or grade) of membership of y in F. In a fuzzy set, the transition between membership and non-membership is gradual rather than abrupt. The universal set U is not fuzzy.

86

The union of two fuzzy subsets F_1 and F_2 of U can be defined as

$$f_{F1 \cup F2} (\gamma) = \max [f_{FA} (\gamma), f_{F2} (\gamma)] \qquad (5.3)$$

The intersection of two fuzzy subsets F1 and F2 of U can be defined as

$$f_{F1 \cap F2} (\gamma) = \min [f_{FA} (\gamma), f_{F2} (\gamma)] \qquad (5.4)$$

where f_{F1} and f_{F2} are membership functions of the fuzzy subsets F_1 and F_2, respectively, and γ is an element in U.

c. Fuzzy random variable

Let Γ be the collection or space of all fuzzy subsets and suppose $(F_1, \ldots, F_n)$ are n unrelated fuzzy subsets (or variables) on Γ. Let γ be an element in Γ; then the membership of γ in a fuzzy subset F_i is determined by the membership function $f_{F1} (\gamma)$ of F_i. Here $f_{F1} (\gamma)$ is the degree of membership of γ in F_i. Suppose that Ω is a probability space. Further, suppose $A_1 \ldots A_n$ are the events that are the subsets of the probability space Ω.

Let X_{Ai} be a random variable associated with event A_i, and each outcome $\omega \in A_i$ in Ω specifies the value of X_{Ai}, that is, $X_{Ai} (\omega)$. The random variable X_{Ai} can be described by any discrete probability distribution to provide the value of $X_{Ai} (\omega)$. Then a fuzzy random variable Y is a function from the probability space Ω into a space of fuzzy subsets Γ; that is, $Y: \Omega \rightarrow \Gamma$. In other words, Y can correspond to any one of n possible events (i.e., $A_1 \ldots A_n$) corresponding to the fuzzy subset (i.e., F_1, $\ldots$, F_n) and can have a value from the membership range of the membership function (i.e., $f_{F1}, \ldots, f_{Fn}$).

If [0, 1] is considered as a non-suitable membership scale for a fuzzy variable F_i, the points on the real line correspond to the range (a specified subinterval within the interval [0, 1]) of a membership function f_{Fi} rather than the domain [0, 1].

5.3 Fuzzy Logic based Modeling of Solar Irradiance

In this section, seven Indian stations have been considered on the basis of their climatic conditions. In India some climatic parameters like solar irradiance, sunshine hours, temperature, rainfall, atmospheric pressure and humidity are measured by Indian Meteorological Department. However, these parameters are not being measured particularly, solar irradiance at all stations. Most of the stations are relatively new and therefore have not accumulated long-term data. In the present model, only two parameters that can be easily measured have been considered as inputs and remaining parameters are assumed to be constant. However each parameter has their own effects on the solar irradiance and their negligence introduces some error in the estimation. Hence, in this analysis a simple technique is presented, where all the uncertainties and model complications related to solar irradiance and sunshine duration and temperature are treated linguistically using fuzzy sets. In the fuzzy approach there is neither any assumption nor any model constant. Therefore this method could minimize the errors in the estimation of solar irradiance. The MATLAB Toolbox (FIS) is used to relate the solar irradiance with sunshine duration and temperature. The inputs for all the above mentioned stations are fuzzified into five fuzzy subsets such as very low, low, medium, high and very high values of input data. However all the data used in this analysis is expressed in the normalized form. Similarly the solar irradiance values are also subdivided into five groups according to increasing magnitude.

The following methodology has been adopted in the development of fuzzy based model for solar energy estimation in MATLAB toolbox. The fuzzy inference system editor is presented in Figure 5.1.

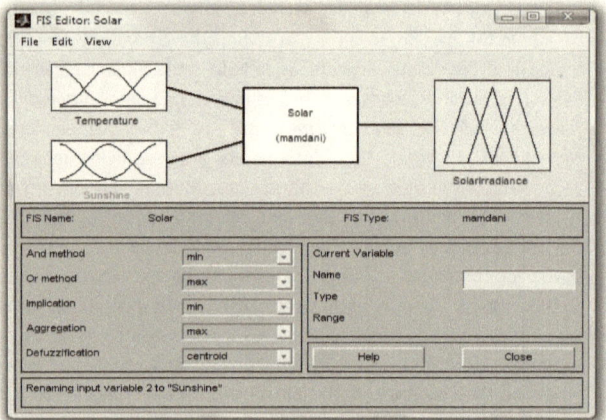

Figure 5.1 Fuzzy inference system editor

The next step is to define membership functions associated with each of the variables. The membership function editor is the tool that display and edit all the membership functions for the entire FIS, including both input and output variables. In the proposed work triangular membership function has been chosen for the input and output variables. The membership functions for sunshine duration, temperature and global irradiance are shown in Figure 5.2, Figure 5.3 and Figure 5.4 respectively.

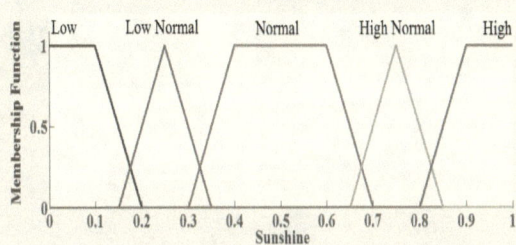

Figure 5.2 Fuzzy subsets membership functions for sunshine duration

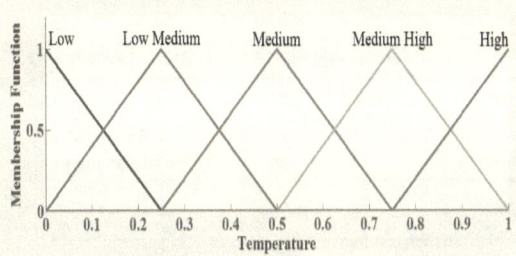

Figure 5.3 Fuzzy subsets membership functions for temperature

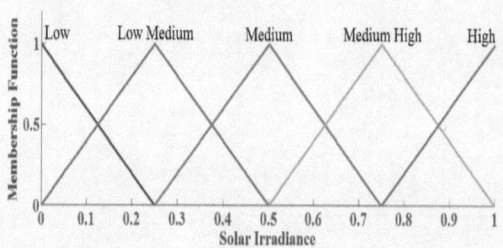

Figure 5.4 Fuzzy subsets membership functions for solar irradiance

All the data used in the proposed work is expressed in the normalized form. Membership functions with their range are presented in Table 5.1.

Table 5.1 Membership function and range of parameters

Inputs				Output	
Membership function	T/T_0	Membership function	S/S_0	Membership function	H/H_0
Low	[0.00-0.25]	Low	[0.00-0.20]	Low	[0.00-0.20]
Low-Med	[0.10-0.50]	Low-Nor	[0.15-0.35]	Low-Nor	[0.15-0.35]
Medium	[0.25-0.75]	Normal	[0.30-0.70]	Normal	[0.30-0.70]
Med-High	[0.50-0.80]	High-Nor	[0.65-0.85]	High-Nor	[0.65-0.85]
High	[0.75-1.00]	High	[0.80-1.00]	High	[0.80-1.00]

For the estimation of solar irradiance at a particular station, a set of multiple-antecedent fuzzy rules have been established. The inputs to the rules are the ratio of sunshine duration and ratio of temperature, and the output consequent is the solar irradiance as given in Table 5.2. The consequents of the rules are in the shaded part of the matrix. The input variables: temperature ratio and sunshine duration ratio are described by the fuzzy linguistic variables such as high, medium-high, medium, low-medium and low and the output variable (solar irradiance) is described as low, low-normal, normal, high-normal and high. To estimate the solar irradiance at any station of India, fuzzy rules have been defined as shown in Figure 5.5.

Table 5.2 Decision matrix for determining global solar irradiance

AND		S_0/S				
		Low	Low-Nor	Nor	High-Nor	High
Temperature	Low	Low-Med	Low-Med	Low	Low-Med	Med
	Low-Med	Medium	Low-Med	Low-Med	Low-Med	Med
	Medium	High-Med	Med	Low-Med	Low-Med	High-Med
	Med-High	High-Med	High-Med	Med	High-Med	High-Med
	High	High	High-Med	Med	High-Med	High

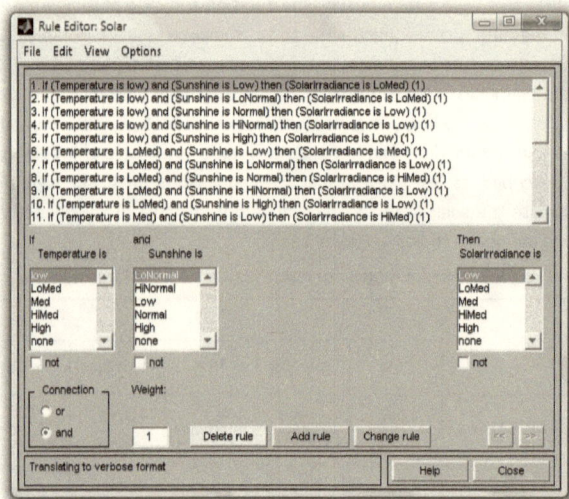

Figure 5.5 Fuzzy rules

The rule viewer as shown in Figure 5.6 presents a sort of micro view of the fuzzy inference system by showing one calculation at a time in detail.

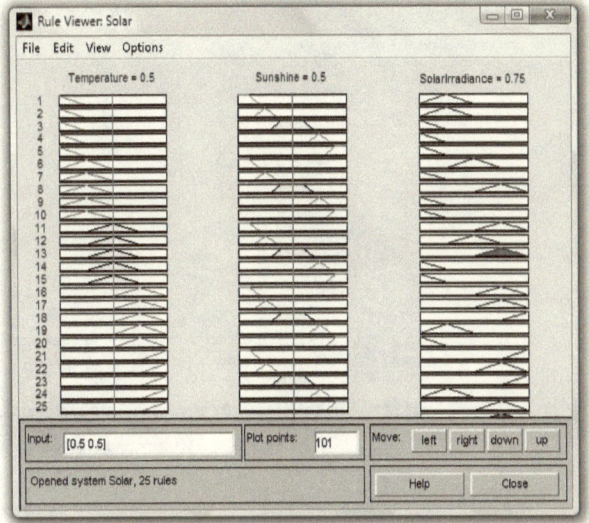

Figure 5.6 Fuzzy rules viewer

The surface viewer presents the entire output surface of the system, based on the entire span of the input set. Three dimensional surfaces of output of fuzzy model is described in Figure 5.7.

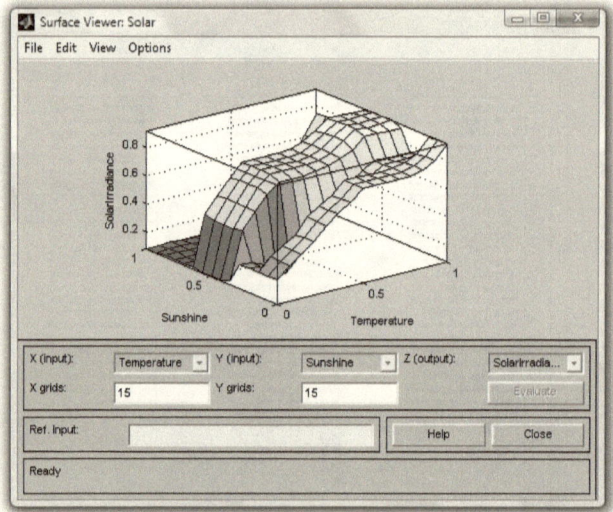

Figure 5.7 Three dimensional surfaces of output of fuzzy model

5.4 Artificial Neural Network

In the previous section, an intelligent model based on Fuzzy Logic technique for solar energy estimation has been presented. This technique has taken into account the uncertainties involved in the atmosphere in different weather conditions. The performance of the developed model is found better as compared to the mathematical and regression model. Due to nonlinear relationship between input and output parameters the percentage RMSE in the estimation of global solar irradiance quite more using fuzzy model. An artificial neural network provides computationally efficient way of determining an empirical, possibly nonlinear relationship between a number of inputs and one or more outputs. ANN has been applied for modeling identification, optimization, prediction, forecasting, designing, sizing and performance of energy and renewable energy systems and control of complex system.

In this section, an artificial neural network (ANN) based model is presented for the assessment of solar energy at any given location in India based on its latitude, longitude, altitude, sunshine hours, temperature and the month of the year. However, the parameters like latitude, longitude and altitude are constant for particular location.

5.5 Artificial Neural Network for Solar Energy Assessment

The model of an artificial neuron is shown in Figure 5.8. In this model the processing elements (neuron) computes the weighted sum of its inputs and outputs according to whether this weighted input sum is above or below a certain threshold θ_k. The externally applied bias has the effect of lowering the net input of the activation function.

$$y_k = f(u_k\_\theta_k) \qquad\qquad (5.5)$$

$$u_k = \sum w_{kj}x_j \qquad\qquad (5.6)$$

Here x_1, x_2 ...x_p are input signals and w_{k1}, w_{k2} ...w_{kp} are interconnection weights of the neuron k. u_k is the linearly combined output.

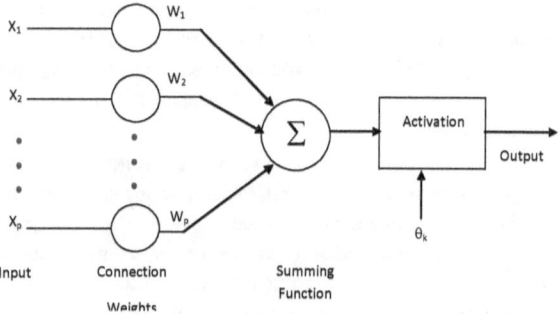

Figure 5.8 (a) Artificial Neuron Model

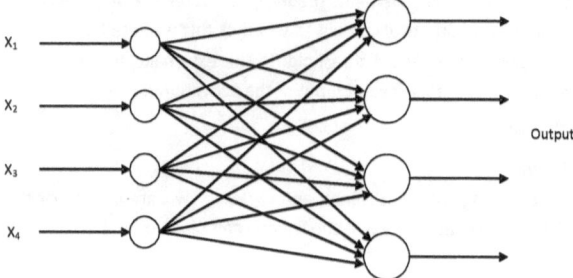

Figure 5.8 (b) Single Layer Feed Forward Network

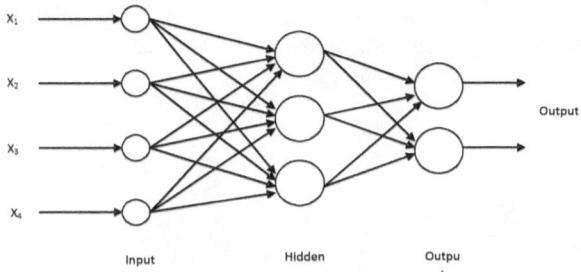

Figure 5.8 (c) Multi-Layer Feed Forward Network

The interconnection of several layers forms a multilayer feed-forward networks as shown in Figure 5.8 (c). The layer between the input and output layer is called a hidden layer and its function is to intervene between the external input and output. The hidden layer had no direct contact with the external environment. The radial basis function network is a special class of multilayer feed-forward networks. The hidden layer in this employs a radial basis function, such as a Gaussian kernel as the activation function. Feedback networks that have closed loops are called recurrent networks. In single layer recurrent network the processing element output is feedback to itself or to other processing element or to both. In the multiplayer recurrent network the neuron output can be directed back to the nodes in the preceding layers.

The architecture used in this work is shown in Figure 5.9 has an input layer of six inputs, one hidden layer with a tan-sigmoid activation function, ϕ, defined as

$$\phi = 1/(1 + e^{-n}) \qquad (5.7)$$

Where n is the corresponding input.

The MATLAB neural network toolbox is used for the implementation of the feed forward back propagation network. The flowchart for the same algorithm is shown in Figure 5.10.

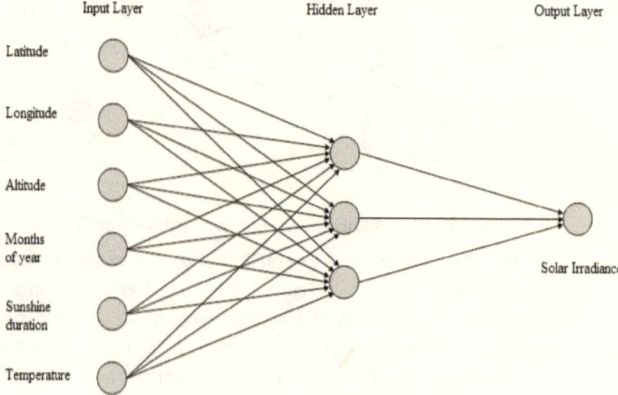

Figure 5.9 General ANN architecture

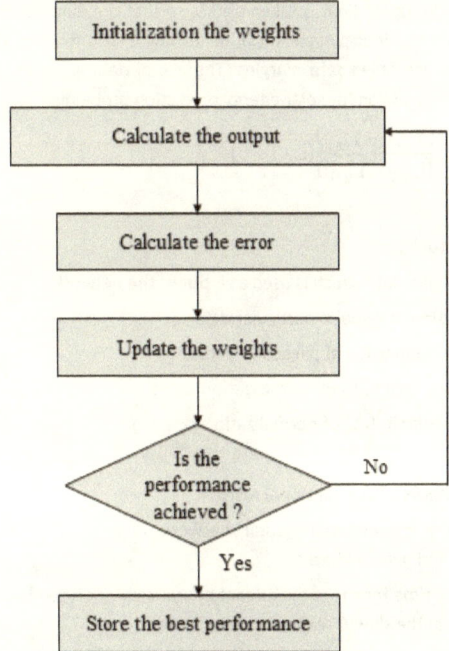

Figure 5.10 Flowchart for ANN model

5.6 Normalization of Meteorological Data for Solar Energy Assessment

     The input and output data for the neural network may have different ranges if actual monthly data is directly used. This may cause convergence problem during the learning process. To avoid such problem, the input and output data are scaled such that they remain within the range of (0.1-0.9). The lower limit is 0.1, so that during testing it could not go far beyond lower extreme limit, which is 0.

Similarly, the upper limit is taken as 0.9, so that the data could go upto upper extreme limit, which is 1.0, in testing. These margins of 0.1 on both sides are called safe margins. The actual data is scaled using the following expression for solar energy prediction problem:

$$L_S = \frac{\left(Y_{max} - Y_{min}\right)}{\left(L_{max} - L_{min}\right)}\left(L - L_{min}\right) + Y_{min} \qquad (5.8)$$

where

L = The actual data

L_s = The scaled data which is used as input to the network

L_{max} = Maximum value of particular data set

L_{min} = Minimum value of particular data set

Y_{max} = Upper limit (0.9) of normalization range

Y_{min} = Lower limit (0.1) of normalization range

5.7 Drawbacks of Conventional ANN

The conventional neural network model suffers from the following serious drawbacks.

1. Training time for the conventional neural network is too large, which results in the slower response of the system.

2. Number of hidden layers and hidden neurons can't be predicted accurately, and also they are large in number for complex function approximation.

3. The existing neuron model performs only the operation of summation of its weighted inputs; it does not perform the operation of product on its weighted inputs.

4. There is a effect of threshold (activation) function on training time also, accuracy of test result depends on threshold function.

5. Back propagation learning also has some shortcomings, like: Slow learning and problem of local minima may occur in the system.

6. There is an effect of normalization range of training data. Hence selection of suitable range (i.e. maximum and minimum values) is of great importance as it affects the results of the neural network training.

7. Training time of neural network depends on the mapping of input-output pattern (I/O-mapping) presented to the network.

8. Training time of the network also depends on the sequence of presentation of data.

To overcome the above drawbacks number of variants has been developed in the past decades. Most of the variants are either burden on the learning algorithms or/and increases the computational labour. In this thesis, a new generalized neuron model is used, which overcome the drawbacks of conventional neural network by performing various possible variations and modification in the previous model for the problem of solar energy prediction. The model should incorporate non-linearities present in the system. The model should also incorporate following features:

1. The generalized neural network should consist of characteristics of simple neuron and also high order neuron characteristics.

2. There is no need of the selection of number of hidden layers and the number of neurons i.e. the complexity of the network should reduce.

3. The input output mapping should not affect the response of the network.

4. Normalizing effect should not be there.

5.8 Development of Generalized Neural Model

The general structure of the common neuron is an aggregation function and its transformation through a filter. It is shown in the literature that the ANNs can be universal function approximators for given input-output data. The common neuron structure as shown in Figure 5.11 has summation as the aggregation function with sigmoidal, radial basis, tangent hyperbolic or linear limiters as the thresholding function. The aggregation operators used in the neurons are generally crisp. However, they overlook the fact that most of the processing in the neural networks is done with incomplete information at hand. Thus, a GN model approach has been adopted that uses the fuzzy compensatory operators that are partly sum and partly product to take into account the vagueness involved.

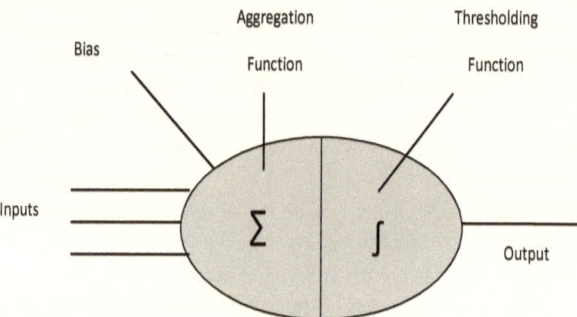

Figure 5.11 Simple Neuron Model

Use of the sigmoidal thresholding function and ordinary summation or product as aggregation function in the existing model fails to cope up with the nonlinearities involved in real life problems. To deal with these, the proposed model has both sigmoidal and gaussian functions with weight sharing. The GNN model has flexibility at both the aggregation and threshold function level to cope up with the nonlinearity involved in the type of applications dealt with, as shown in Figure 5.12.

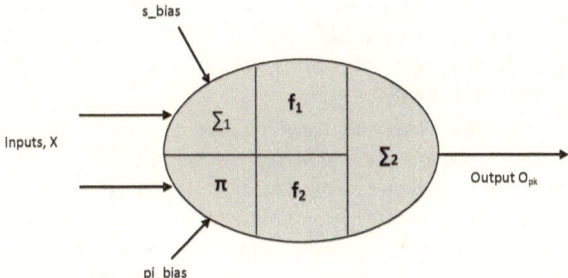

Figure 5.12 Generalized Neuron Model

The neuron model described above is known as the summation type compensatory neuron model, since the outputs of the sigmoidal and Gaussian functions are summed up. Similarly, the product-type compensatory neuron models may also be developed. It is found that in most of the applications, summation-type compensatory neuron model works well.

5.9 GNN Model for Solar Energy Estimation

Existing models of neuron in the structure of artificial neural network use the sigmoidal activation function and ordinary summation as aggregation functions. These models face problems in training when non-linearity involved in real life problems. To deal with the above, the proposed model has both summation (Σ) and product (π) as aggregation function. The generalized neuron model have flexibility at both the aggregation and activation function level to cope up with the non-linearity involved in the type of applications dealt with. The product and power non-linearity in problems made them complex for training, but with the help of product aggregation function it is quite easy to train. In this chapter we have tried product neuron layers along with summation neuron layers in ANN and found that the training time is drastically reduced for mapping the non-linear system, if one layer is product and other is summation layer.

The generalized neural network is developed on the basis of the Boolean algebra. It is the well-known that with the help of sum of product and product of sum one can implement any given function. Similarly in the generalized neuron structure summation and product as aggregation functions have been incorporated and the aggregated outputs pass through a non-linear squashing / thresholding function as shown in the Figure 5.13. Σ-part have the summation of weighted input with sigmoidal activation function f_1, while the part have the product of weighted input with Gaussian activation function f_2. The final output of the neuron is a function of the weighted outputs O_Σ and O_π.

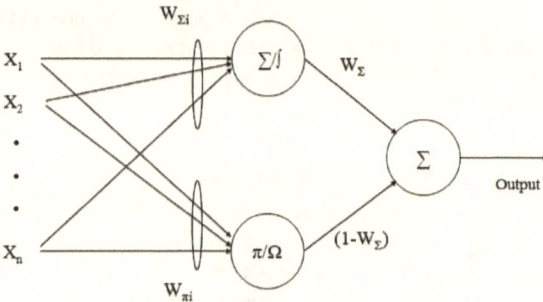

Figure 5.13 Generalized Neuron Model

The output of summation (Σ) part of the generalized neuron is

$$O_\Sigma = f_1 (\Sigma W_{\Sigma i} X_i + X_{o\,\Sigma}) \qquad\qquad (5.9)$$

The output of product (π) part of the generalized neuron is

$$O_\pi = f_2 (\pi W_{\pi i} X_i + X_{o\,\pi}) \qquad\qquad (5.10)$$

Finally, the outputs are summed up to get the neuron output. The output of the neuron can be mathematically written as

$$O_j = O_\Sigma * W_\Sigma + O_\pi (1 - W_\Sigma) \qquad\qquad (5.11)$$

Existing model of neuron in the structure of artificial neural network use the sigmoidal activation function and ordinary summation as aggregation functions. These models face problems in training when non-linearity involved in real life problems.

In this chapter, Fuzzy logic, ANN and GNN techniques have been used for solar energy estimation. However, these techniques can also be used for the prediction of wind speed, weather conditions, SPV system assessment, control of power plants based on renewable energy sources as well as conventional sources etc.

References

1. M. Rizwan, Majid Jamil, Sheeraz Kirmani and D. P. Kothari, "Fuzzy Logic based Modeling and Estimation of Global Solar Energy using Meteorological Parameters", *Energy: The International Journal (Elsevier), USA,* Vol. 70, pp. 685-691, 2014. ISSN: 0360-5442.

2. M. Rizwan, Majid Jamil and D. P. Kothari, "Generalized Neural Network Approach for Global Solar Energy Estimation in India," *IEEE Transactions on Sustainable Energy,* USA, Vol. 3, No. 3, July 2012, pp. 576-584. ISSN: 1949-3029.

3. D. K. Chaturvedi, *"Soft computing techniques and its applications in electrical engineering,"* Springer Verlag Berlin Heidelberg, 2008.

4. L. A. Zadeh, *Fuzzy sets*, Wiley, Newyork, 1992.

5. Zekai Sen, "Fuzzy algorithm for estimation of solar irradiation from sunshine duration," *Solar Energy*, vol. 63, pp. 39-49, 1998.

6. L. A. Zadeh, *Fuzzy sets information control*, 1965.

7. D. Duboi and H. Prade, *Advances in fuzzy set theory and applications*, North-Holland, 1979.

8. A. Kandel, *"Fuzzy mathematical techniques with applications,* Addison-Wesley, 1986.

9. V. Gomez and A. Casanovas, "Fuzzy modeling of solar irradiance on inclined surfaces," *Solar Energy*, vol. 75, pp. 307-315, 2003.

10. N. K. Gautam and N. D. Kaushika, "A model for the estimation of solar radiation using fuzzy random variables," *Journal of Applied Meteorology*, pp. 1267-1276, 2002.

11. M. Pauliscu and P. Gravila, "Fuzzy logic algorithm for atmospheric transmittances of use in solar energy estimation," *Energy Conversion and Management*, vol. 49, pp. 3691-3697, 2008.

12. E. Tulcan-Paulescu and M. Paulescu, "Fuzzy modeling of solar irradiation using air temperature data," *Theoretical and Applied Climatology*, vol. 91, pp. 181-192, 2008.

13. [R. Iqdour and a. Zeroual, "Prediction of daily global solar radiation using fuzzy systems," *International Journal of Sustainable Energy*, vol. 26, pp. 19-29, 2007.

14. D. I. Tseles, A. I. Dounis and J. Zisos, "Meteorological parameters forecasting for renewable energy systems using soft computing techniques," *International Journal of Sustainable Energy*, vol. 26, pp. 30-36, 2007.

15. F. R. Rubio, M. Berenguel and E. F. Camacho, "Fuzzy logic control of solar power plant," *IEEE Transactions on Fuzzy Systems*, vol. 3, pp. 459-467, 1995.

16. Haykin, *Artificial neural network*, PHI, India, 2007.

17. S. Rehman and M. Mohandes, "Artificial neural network estimation of global solar radiation using air temperature and relative humidity," *Energy Policy*, vol. 36, pp. 571-576, 2008.

18. M. Mohandes, S. Rehman and T. O. Halawa, "Use of radial basis functions for estimating monthly mean daily solar radiation, *Solar Energy*, vol. 68, pp. 161-168, 2000.

19. S. M. Al-Alawi and H. A. Al-Hinai, "An ANN based approach for predicting global radiation in locations with no measurement instrumentation," *Renewable Energy*, vol. 14, pp. 199-204, 1998.

20. T. Krishnaiah, S. S. Rao, K. Madhumurthy and K. S. Reddy, "Solar global radiation prediction for India using ANN model," *Proceedings of International Conference on Solar Radiation and Day Lighting(SOLARIS)*, New Delhi, pp. 91-98, 2007.

21. A. Sozen, E. Arcaklioglu and M. Ozalp, "Estimation of solar potential in Turkey by artificial neural network using meteorological and geographical data," *Energy Conversion & Management*, vol. 45, pp. 3033-3052, 2004.

22. M. Benghanem, A. Mellit and S. L. Alamri, "ANN based modeling and estimation of daily global solar radiation data: A case study," *Energy Conversion and Management*, vol. 50, pp. 1644-1655, 2009.

23. J. L. Bosch, G. Lopez and F. J. Batlles, "Daily solar irradiation estimation over a mountainous area using artificial neural network," *Renewable Energy*, vol. 33, pp. 1622-1628, 2008.

24. K. F. Dagestad and J. A. Olseth, "A modified algorithm for calculating the cloud index," *Solar Energy*, vol. 81, pp. 280-289, 2007.

25. A. Mellit, "Artificial intelligence technique for modeling and forecasting of solar radiation data: A review," *International Journal on Artificial Intelligence and Soft Computing*, vol. 1, pp. 52-76, 2008.

26. A. Mellit, M. Benghanem and M. Bendekhis, "Artificial neural network model for prediction solar radiation data: Application of sizing standalone photovoltaic power system," *Proceedings of IEEE Conference*, pp. 1-5, 2005.

27. O. Senkal and T. Kulali, "Estimation of solar radiation over Turkey using artificial neural network and satellite data," *Applied Energy*, vol. 77 pp. 1-7, 2008.

28. A. S. S. Dorlvo, J. A. Jervase and A. A. Lawati, "Solar radiation estimation using artificial neural network," *Applied Energy*, vol. 71, pp. 307-319, 2002.

29. A. Mishra, N. D. Kaushika, G. Zhang and J. Zhou, "Artificial neural network model for the estimation of direct solar radiation in Indian zone," *International Journal of Sustainable Energy*, vol. 23, pp. 95-103, 2008.

30. G. Lopez, F. J. Batlles and J. T. Pescador, "Selection of input parameters to model direct solar irradiance by using artificial neural network," *Energy*, vol. 30, pp. 1675-1684, 2005.

31. S. Abe and M. S. Lan, "Fuzzy rules extraction directly from numerical data for function approximation," *IEEE Transactions on Systems, Man and Cybernetics*, vol. 25, pp. 289-302, 1995.

32. T. Krishnaiah, S. S. Rao and K. S. Reddy, "Neural network approach for modeling global solar radiation," *Journal of Applied Sciences Research*, vol. 3, pp. 1105-1111, 2007.

33. A. Mellit, "Sizing of photovoltaic systems: A review," *Renewable Energy*, vol. 10, pp. 463-472, 2007.
34. A. Mellit, M. Benghanem and S. A. Kalogirou, "Modeling and simulation of a stand-alone photovoltaic system using an adaptive artificial neural network: Proposition for a new sizing procedure," *Renewable Energy*, vol. 32, pp. 285-313, 2007.
35. A. Mellit, S. A. Kalogirou and M. Drif, "Application of neural networks and genetic algorithm for sizing a photovoltaic systems," *Renewable Energy*, vol. 35, pp. 2881-2893, 2010.
36. A. Azadeh, A. Maghsoudi and S. S. Khani, "Using an integrated artificial neural network model for predicting global radiation: The case study of Iran," *Energy Conversion and Management*, vol. 50, pp. 1497-1505, 2009.
37. A. Chaouachi, R. M. Kamel , H. Hayashi and K. Nagasaka, " Neural network ensemble based solar power generation short term forecasting," *World Academy of Science, Engineering & Technology*, vol. 54, pp. 5459-5465, 2009.
38. A. Mellit, "Artificial intelligence technique for modeling and forecasting of solar radiation data: A review," *International Journal on Artificial Intelligence and Soft Computing*, vol. 1, pp. 52-76, 2008.
39. A. Mellit, M. Benghanem and S. A. Kalogirou, "Modeling and simulation of a stand-alone photovoltaic system using an adaptive artificial neural network: Proposition for a new sizing procedure," *Renewable Energy*, vol. 32, pp. 285-313, 2007.
40. M. C. Mabel and E. Fernandez, "Analysis of wind power generation and prediction using ANN: A case study," *Renewable Energy*, vol. 33, pp. 986-992, 2008.
41. D. K. Chaturvedi, M. Mohan and P. K. Kalra, "Improved generalized neuron model   for short-term load forecasting," *Soft Computing*, Vol. 8, pp. 370-379, 2004.

42. D. K. Chaturvedi, P. S. Satsangi and P. K. Kalra, "Fuzzified neural network approach for load forecasting," *Engineering Intelligent Systems*, vol.1, pp. 3-9, 2001.
43. Manmohan, D. K. Chaturvedi, P. S. Satsangi and P. K. Kalra, "Neuro fuzzy approach for development of new neuron model," *Soft Computing*, vol. 8, pp. 19-27, 2003.
44. D. K. Chaturvedi, O. P. Malik, and P. K. Kalra, "Performance of a generalized neuron-based PSS in a multimachine power system." *IEEE Transactions on Energy Conversion*, vol. 19, pp. 625-632, 2004.
45. D. K. Chaturvedi, P. S. Satsangi and P. K. Kalra, "Load frequency control: a generalized neural network approach," *International Journal of Electrical Power & Energy Systems,* vol. 21, pp. 405-415, 1999.
46. D. K. Chaturvedi, Sinha Anand Premdayal and Ashish Chandiok, "Short-term load forecasting using soft computing techniques," *International Journals of Communication Networks and System Sciences*, vol. 3, pp. 273-279, 2010.
47. D. K. Chaturvedi, R. Chauhan and P. K. Kalra, "Applications of generalized neural networks in for aircraft landing control system," *Soft Computing*, vol. 6, pp. 441-448, 2002.

Chapter 6
Distribution Systems and Recative Power
Control introduction

Sheeraz Kirmani

6.1 Introduction

Distribution system differs from transmission system in several ways such as the number of branches and sources is much higher in distribution networks and the general structure and topology is quite different. A typical system consists of a step-down (e.g. 66/11KV) on load tap changing transformer at a bulk supply point feeding a number of cables of different lengths. A series of step down 3-phase transformers, e.g.11KV/415Vare spaced along the routes where the consumers are supplied through 3-phase, 4 wire network giving 400V. While designing a distribution system one has to keep in mind (1) service conditions (2) electrical design (3) mechanical design and (4) various costs.

In service conditions mainly load study is carried out which tells us the type of load to be served, density of consumers etc. The Main loads are residential, commercial, industrial, municipal traction etc.

The distribution system may be subdivided into primary distribution, distribution transformers, secondary distribution and consumer's service connections. The proper voltage, location sizes and protective equipment must be chosen for various components of distribution system.

Types of distribution system

The electric power today is universally distributed by AC systems. The distribution system mainly consists of following two parts

(1) Primary distribution system, and

(2)The secondary distribution system

Studies have indicated that as much as 13% of total power generated is consumed as I^2R losses at the distribution level [1]. Reactive currents account for a portion of these losses. However, the losses produced by reactive currents can be reduced by the installation of shunt capacitors. In addition to the reduction of energy and peak power losses, effective capacitor installation can also release additional kVA capacity from distribution apparatus and improve the system voltage profile. Thus, the problem of optimal capacitor allocation involves determining the locations, sizes, and number of capacitors to install in a distribution system such that the maximum benefits are achieved while all operational constraints are satisfied at different loading levels. Published literature describing capacitor placement algorithms are abundant. The Capacitor Subcommittee of the IEEE Transmission and Distribution Committee has published 10 bibliographies on power capacitors from 1950 to 1980 [2]. Moreover, the VAR Management Working Group of the IEEE System Control Subcommittee has published another bibliography on reactive power and voltage control in power systems [3]. The total publication count listed in these bibliographies is over 400, and many of these papers are specific to the problem of optimal capacitor allocation. Therefore, to survey all capacitor placement literature would be an enormous task.

6.2 Types of Distribution systems

Three are three different ways to lay out a power distribution system used by electric utilities. The radial, loop and network systems differ how the distribution feeders are arranged and interconnected about a substation.

Most power distribution systems are designed as radial distribution systems, the radial system is characterized by having only one path between each customer and a substation The electrical power flows exclusively away from a substation and out to a consumer along a single path, which if interrupted results in complete loss of power to the consumer

Because load and power factors are easy to establish using voltage profiles, therefore they can be determined with a good degree of accuracy without resorting to exotic calculation methods. Further equipment capacity requirement can be ascertained exactly, fault levels can be predicted with a reasonable degree of accuracy and protective devices breaker relays and fuses can be coordinated in an absolutely assured manner without resorting to network methods of analysis. Regulators and capacitors can be sized, located and set using relatively simple procedures.

On the debit side, radial feeder systems are less reliable than loop or network systems because there is only one path between the substation and customer, thus if any element along this path fails, a loss of power delivery results. Generally when such a failure occurs a repair crew is dispatched to re switch temporarily the radial pattern network, transferring the interrupted consumer to another feeder until the damaged element can be repaired. This minimizes overall outage but an overall outage still occurred because of failure.

Despite this apparent flaw, radial distribution systems if well designed and constructed generally provides a very high level of reliability. For all but the most densely populated areas or absolutely critical loads the additional cost of an inherently more reliable configuration (loop or network) cannot possibly be justified for a slight improvement that is gained over a well-designed radial system.

An alternative to purely radial feeder design is the loop design consisting of a distribution design with two paths between the power sources and every consumer. Equipment is sized and each loop is designed so that service can be maintained regardless of where an open point could be on the loop. Because of this requirement whether operated radially (with one open point in each loop) or with closed loops, the basic equipment capacity requirements of the loop feeder design do not change.

In terms of complexity a loop feeder system is slightly more complex than a radial system and power usually flows out from both sides towards the middle and in all cases takes only one of the two routes. Voltage drop sizing and protection engineering are slightly more complicated than for radial systems.

The major disadvantage of loop systems is capacity and cost. A loop must be able to meet all power and voltage drop requirements from only one end but not from both. It needs extra capacity on each end, and the conductor must be large enough to meet the power and voltage drop needs of the entire feeder if feed from either end. This makes loop system inherently more reliable than a radial system, but the larger conductor and extra capacity result in increased cost.

Distribution networks are the most complicated, most reliable, and in very rare cases the most economical method of distributing electrical power. Network involves multiple paths between all points in the network. Power flow between any two points is usually split among several paths, and if a failure occurs it instantly and automatically re-routes itself. Networks are more expensive than radial distribution systems, but not greatly so in some instances .In dense urban population where the load density is very high, the distribution must be placed underground, and also where repair and maintenance is difficult because of traffic and congestion, network may cost little less than loop systems. Networks require only a little more conductor capacity than a loop system. The loop configuration requires "double capacity "everywhere to provide increased reliability.

Networks have the major disadvantage, they are much more complicated than other forms of distribution and thus much more difficult to analyze. Loadings and power flow, faults currents and protection must be determined by network techniques such as those uses by transmission planners.

As mentioned above most of the power distribution systems are designed as radial distribution systems so I have taken radial system for my analysis of reactive power compensation.

6.3 Transfer of power between active sources

If a line of series inductive reactance X and series resistance R is considered, then

$$Q_{Loss}=XI^2 \ =X\ (P_R{}^2+Q_R{}^2)/E_R{}^2 \qquad\qquad (1.1)$$

$$P_{Loss}=RI^2=R\ (P_R{}^2+Q_R{}^2)/E_R{}^2 \qquad\qquad (1.2)$$

Where P_R and Q_R are the active and reactive power at the receiving and respectively and E_R is the voltage at the receiving end.

We see from the above equation that the reactive power absorbed by X is XI^2

A companion term to active power loss RI^2 associated with the resistive elements. As seen from the above equation that an increase of reactive power transmitted increases the active power as well as the reactive power losses. This has an impact on efficiency of power transmission and voltage regulation from the above we can draw the following conclusions

116

- Active power transfer depends mainly on the angle by which the sending end voltage leads the receiving end voltage.

- Reactive power transfer depends mainly on voltage magnitudes .It is transmitted from the side of higher voltage magnitude to the side of lower voltage magnitude.

- Reactive power cannot be transmitted over long distances since it would require a large voltage gradient to do so.

- An increase in reactive power transfer causes an increase in active as well as reactive power losses.

For efficient and reliable operation of power systems, the control of voltage and reactive power should satisfy the following requirement

- Voltage at terminals of all equipment's is within the acceptable limits. Both utility equipment and customer equipment are designed to operate at a certain voltage rating. Prolonged operation of system equipment at voltages outside the allowable range could adversely affect their performance and possibly can cause them damage.

- The reactive power flow is minimized so as to reduce RI^2 and XI^2 losses to a practical minimum. This ensures that the system operates efficiently i.e. mainly for active power transfer.

6.4 Production and absorption of reactive power

As loads vary the reactive power requirements of the system also vary. Since reactive power cannot be transmitted over long distances, therefore voltage control has to be effected by using special devices dispersed throughout the system. The proper selection and coordination of of equipment for controlling reactive power and voltages is the major challenge for power system engineer. Following are the devices used to generate or absorb reactive power:

Synchronous generators can generate or absorb reactive power depending on the excitation. When over excited they supply the reactive power, and when under excited they absorb the reactive power.

Overhead lines, depending on the load current, either absorbs or supply the reactive power at loads below the natural load the lines produced the net reactive power; at loads above the natural loads they absorb the reactive power.

Underground cables, owing to their high capacitance, have high natural loads and hence generate reactive power under all conditions.

Transformers always absorb the reactive power regardless of their loadings. At no load the shunt magnetizing reactance effect is predominate, and at full load series leakage inductive effect is predominate.

Loads normally absorb reactive power. A typical load bus supplied by power system composed of a large number of devices. The composition changes depending on the day, season and weather conditions .The composite characteristics are such that the load bus absorbs the reactive power .Both active and reactive power of a composite load varies as a function of voltage magnitudes. Loads at lagging power factor cause excessive voltage drops in the system and are uneconomical to supply.

Compensating devices are usually added to supply or absorb the reactive power and there by control the reactive power balance in a desired manner.

6.4.1 Methods of voltage control

The control of voltage levels is accomplished by controlling the production, absorption and flow of reactive power at all levels in the system The devices used for controlling the voltage profile throughout the system are

- Sources and sinks of reactive power, such as shunt capacitors, shunt reactors, synchronous condensers and static var compensators.

- Line reactance compensators, such as series capacitors.

- Regulating transformers, such as tap changing transformers and boosters.

Shunt capacitors and reactors and series capacitors provide passive compensation. They are either permanently connected to transmission and distribution system or switched. They contribute to voltage control by modifying the characteristics.

Synchronous condensers and SVCs provide active compensation, the reactive power absorbed/supplied by them is automatically adjusted so as to maintain the voltages of the buses to which they are connected. Together with the generating units they establish the voltages at specific points in the system. Voltages at other locations in the system are determined by active and reactive power flows through various circuit elements including the passive compensating devices.

6.4.2 Shunt Capacitors

Shunt capacitor supply reactive power and boost the local voltages. They are used throughout the system and are supplied in a wide range of sizes. They are very economical means of supplying the reactive power. The principal advantages of shunt capacitor are their low cost and their flexibility of installation and operation. They are readily applied at various points in the system, there by contributing to efficiency of power transmission and distribution system. The principal disadvantage of shunt capacitor is that their reactive power output is proportional to square of voltage. Consequently the reactive power output is reduced at low voltages when it is likely to be needed most.

6.5 Application to Distribution systems

The objective of capacitor placement is to provide reactive power close to the point where it is being consumed, rather than supply it from long distance. Most loads absorb reactive power i.e. they have lagging power factors. When reactive power is provided only by power plants each system component, e.g. generators, transformers, transmission and distribution lines, switchgear and protective equipment's has to be increased in size. Capacitors can mitigate these conditions by decreasing the reactive power demand all the way back to the generators. Line currents are also reduced from capacitor location all the way back to the generators. As a result losses and loadings are reduced in distribution lines, substation transformers and transmission lines. Installation of capacitors can also increase generator and substation capability to for additional loads.

In general the benefits derived from capacitor installation are:

- Released generation capacity
- Released transmission capacity
- Released distribution substation capacity
- Additional advantages in distribution system are:
- Reduced energy losses
- Reduced voltage drop and hence improve voltage regulation
- Released capacity of feeder and associated apparatus\us
- Revenue increase due to voltage improvement
- Postponement or elimination of capital expenditure due to system
- improvements and/or expenditure

Chapter 7
Optimal Reclosing Techniques for Power Quality
Enhancement

Sagnika Ghosh and Mohd. Hasan Ali

7.1. Introduction

In a power system, auto-reclosing schemes are widely applied on high-voltage transmission lines. It is well realized that the transient faults which are most frequent in occurrence do not cause permanent damage to the system as they are transitory in nature. These faults disappear if the line is disconnected from the system momentarily in order to allow the arc to extinguish. After the arc path has become sufficiently deionized, the line can be reclosed to restore normal service.

If the fault is permanent, reclosing is of no use, as the fault still remains on reclosing. It shows that nearly 80% of the faults are cleared after the first trip, 10% stay in for the second reclosure which is made after a time delay, 3% require the third reclosure and about 7% are permanent faults which are not cleared and result in lockout of the reclosing relay .

The common practice is to reclose the circuit breakers automatically to improve service continuity. The reclosure may be either high-speed or with time delay. High-speed reclosure refers to the closing of circuit breakers after a time just long enough to permit fault-arc de-ionization. The reclosure can be completed in less than 1 second. However, high-speed reclosure is not always acceptable. Reclosure into a permanent fault i.e. unsuccessful reclosure may cause system instability and may deteriorate power quality. Thus, the application of automatic reclosing is usually constrained by the possibility of a persistent fault, which would create a second fault after reclosure.

Conventional auto-reclosing techniques adopt fixed time interval reclosing, that is, the circuit breakers reclose after a prescribed dead time which is set to a constant value. Since the transient stability is dependent on the generator state of reclosing instance, in some cases, conventional method may cause an unstable state, especially for the case of unsuccessful reclosing. Therefore, in order to maintain the synchronism, power quality and enhance the transient stability, the circuit breakers should be reclosed at optimal reclosing time (ORCT) where system disturbances after reclosing operation are restrained effectively.

In literatures, different types of reclosing techniques are reported, such as the Total load angle method, Total kinetic energy method, Variable dead time control method, Variable dead time auto reclosing using artificial neural network, Adaptive reclosing algorithm considering distributed generation, Transient energy function method and Neural-network based adaptive single–pole autoreclosure. The aim of all of these methods is to determine ORCT.

However, if the mentioned methods are summarized, then it is seen that mainly three types of reclosing techniques are adopted, such as the kinetic energy or transient energy based method, variable dead time based method and the adaptive reclosing method considering distributed generation. Although the individual techniques are well documented in the literature, no comparison is made among the methods.

This chapter makes a comparison among the Total load angle method, Total kinetic energy method, and Variable dead time methods, and this is the main contribution of this work. Another point to note here is that the performance of these three methods is compared with that of the conventional reclosing method. The adaptive reclosing method considering distributed generation is not considered in this work. The comparison among the reclosing techniques is done in terms of power quality enhancement of a multi-machine power system. Two indices are used to evaluate the power quality of the system. The proposed research provides an in-depth idea about the choice of reclosing methods.

In order to analyze the optimal reclosing techniques, the IEEE nine bus power system model is considered. Both balanced and unbalanced temporary and permanent faults at different locations of the system are considered. Simulations are performed through Matlab/Simulink software.

The organization of this chapter is as follows. Section 2 explains the basics of circuit breakers. Section 3 describes the model systems for the proposed study. Section 4 explains the optimal reclosing techniques. Section 5 describes the simulation results. Section 6 describes the practicality of the optimal reclosing techniques. Finally, section 7 provides some conclusions regarding this work.

7.2. Description of Model System

For the analysis of optimal reclosing techniques, the IEEE nine bus power system model [4] shown in Fig. 1 is used.

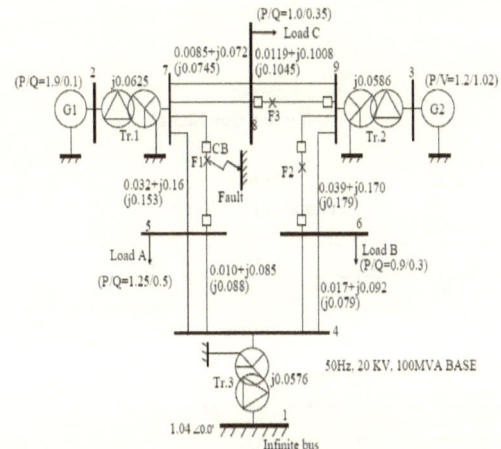

Fig. 1. IEEE 9 bus power system model

The model system consists of two synchronous generators (G1 and G2) with capacities of 200 MVA and 130 MVA, respectively, and an infinite bus. Generators are connected to one another through transformers and double circuit transmission lines. The line parameters have the form R+jX (jB/2), where R, X and B represent resistance, reactance and susceptance per phase with two lines, respectively. F1, F2 and F3 are considered as the fault positions. The AVR (Automatic voltage regulator) and GOV (governor) control models have been considered and shown in Figs. 2(a) and (b), respectively. Parameters of

124

Generator 1 and Generator 2 have been shown in Table I. The system base is 50 Hz and 100 MVA.

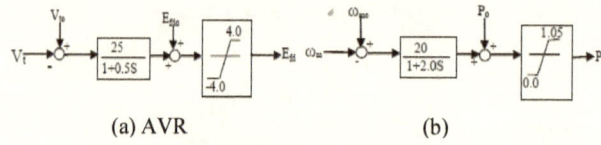

(a) AVR   (b)

Fig. 2. Avr and Gov Models

Table I
Generator Parameters

	G1	G2
MVA	200	130
r_a (pu)	0.003	0.004
x_a (pu)	0.102	0.078
X_d (pu)	1.651	1.220
X_q (pu)	1.590	1.160
X'_d (pu)	0.232	0.174
X'_q (pu)	0.380	0.250
X''_d (pu)	0.171	0.134
X''_q (pu)	0.171	0.134
T'_{do} (pu)	5.900	8.970
T'_{qo} (pu)	0.535	1.500
T''_{do} (pu)	0.033	0.033
T''_{qo} (pu)	0.078	0.141
H (sec)	9.000	6.000

Optimal Reclosing Techniques of Circuit Breakers

Conventional auto-reclosing of circuit breakers can affect the stability and power quality of the system, as it is dependent on the generator state of reclosing instances. So, to enhance the transient stability and power quality, circuit breakers should be closed at an optimal reclosing time, when the system disturbance has no effect after reclosing operation.

As already discussed, there are different types of reclosing techniques. This work focuses on and makes a comparison among three reclosing methods, such as i) Total Load Angle method, ii) Total Kinetic Energy method, and iii) Variable Dead Time method. The following provides an outline of each individual method explaining how the ORCTs are determined.

7.3. Total Load angle Based ORCT Method

In the load angle method, the load angles of both the generators, G1 and G2, are taken into account. The sum of the load angles are considered to find the ORCT [4]. In this method, the optimal reclosing time is considered as the point when the total load angle oscillation of the generators without reclosing operation becomes the minimum first time.

I.   The point when the total load angle oscillation of the generators without reclosing operation becomes the minimum.

II. The value obtained from condition I must be greater than T_r (reclosing time) as shown below:

$$T_r = (10.5 + KV/34.5) \text{ cycles} \qquad (1)$$

where KV indicates the line-to-line rms voltage of the system.

Here in this work, the line-to-line rms voltage is 20 KV. Therefore, the value of T_r is 0.223 sec.

Using this technique the values of ORCT of different fault positions for temporary and permanent faults are calculated from the total load angle responses and represented in the Tables II and III, respectively. In this case also, the ORCT values are certainly bigger than the T_r value in order to allow for the arc to extinguish.

The total load angle response for determining the ORCT at fault position F1 considering both temporary and permanent 3LG (three-phase-to-ground) faults are shown in the Figs. 3.

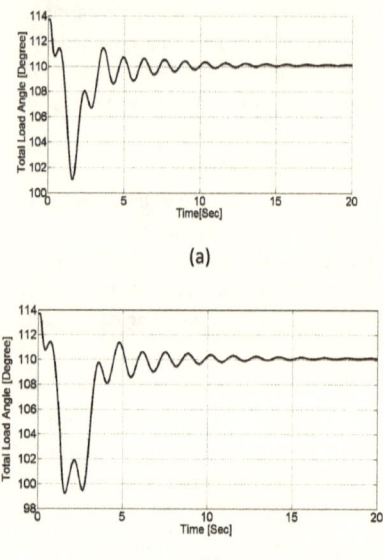

(a)

(b)

Fig.3. Total load angle response for determining ORCT (a) 3LG temporary fault at F1; (b) 3LG permanent fault at F1.

Table Ii

Values of Orct for Temporary Fault

Table Iii

Values of Orct for Permanent Faults

Fault Type	Fault point	Reclosing time (sec) with Conventional method	Reclosing time (sec) with Total Load angle Method	Fault Type	Fault point	Reclosing time (sec) with Conventional method	Reclosing time (sec) with Total Load angle Method
3LG	F1	1.0	1.38	3LG	F1	1.0	0.889
3LG	F2	1.0	1.62	3LG	F2	1.0	0.871
3LG	F3	1.0	1.62	3LG	F3	1.0	0.873
1LG	F1	1.0	1.52	1LG	F1	1.0	0.81
1LG	F2	1.0	1.35	1LG	F2	1.0	0.8
1LG	F3	1.0	1.6	1LG	F3	1.0	0.729

In this method, the total kinetic energy, that is, the sum of the kinetic energies of all the generators is used to determine the ORCT. The total kinetic energy takes not only the rotor speed into account but also the power rating of each generator. In this method, the time when the total kinetic energy oscillation of the generators without reclosing operation becomes the minimum is determined as ORCT. The optimal reclosing time is considered as the point which meets the following conditions:

I. The point when the total kinetic energy oscillation of the generators without reclosing operation becomes the minimum first time.

II. The value obtained from condition I must be greater than T_R(reclosing time). And, this ORCT value must be greater than Tr (reclosing time) as shown in equation (1).

The total kinetic energy, $W_{total,}$ can be calculated easily by knowing the rotor speed of each generator and can be expressed as

$$W_{total} = \sum_{i=1}^{N} W_i (J) \tag{2}$$

where i is the generator number, N is the total number of generators, and

$$W_i = 1/2 * J_i * w_{mi}^2 \, (J) \tag{3}$$

In (3), w_{mi} is the rotor angular velocity in mechanical rad/s and J_i is the moment of inertia in kg.m^2 shown below.

$$J_i = \frac{(H_i \times MVA_{rating_i})}{5.48 \times 10^{-9} \times N_{s_i}^2} \, kg \cdot m^2 \tag{4}$$

In (4), $N_{si} = (120*f) / p$ (5)

where f is system frequency and p is the number of poles.

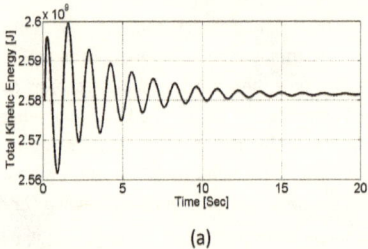

(a)

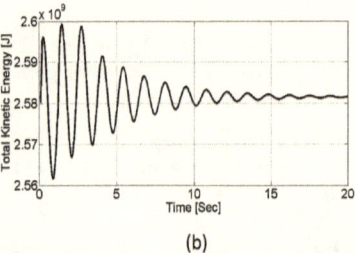

(b)

Fig.4. Total kinetic energy response for determining ORCT (a) 3LG temporary fault at F1; (b) 3LG permanent fault at F1

Using the proposed optimal reclosing technique the values of ORCT of different fault positions for temporary and permanent faults are calculated from the total kinetic energy responses and represented in the Tables IV and V, respectively.

The total kinetic energy response for determining the ORCT at fault position F1 considering both temporary and permanent 3LG (three-phase-to-ground) faults are shown in the Figs. 4.

131

7.4. Variable Dead Time Based ORCT Method

Dead Time in auto reclosing is the time between the opening of circuit breakers and reclosing of circuit breakers to resume transmission of electric power. It means the minimum interval to make auto-reclosing successful due to the deionization of fault arc and the recovery of insulation [6].

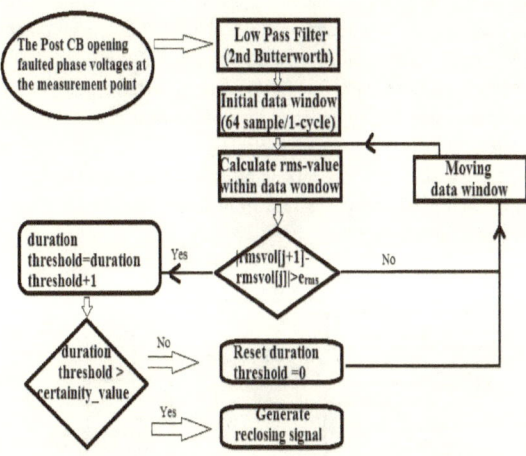

Fig. 5. The block diagram of rms-value tracking method

Fig. 5 shows the rms-value tracking method of the proposed algorithms for deducing arc extinction time, where rmsvol|j| is rms-value of voltage waveform within each data window. €rms is differential threshold that is used for detecting the secondary arc extinction. Duration threshold and certainty value is counter and sample number, respectively, that signify the duration time in order to distinguish complete extinction from re-striking of secondary arc. When difference-value between present rms-value and previous rms-value at each time step, i.e., | rmsvol[j+1] - rmsvol[j] | is greater than or equal to €rms, duration threshold is incremented and as soon as it attains a certainty value, this indicates the secondary arc extinction and a reclosing command signal to excite reclosing relay is generated.

Fig. 6 shows the variable dead time auto-reclosing scheme for hardware implementation. The voltage waveform of the faulted phase at the measurement point can be a digital input to the main process of variable dead time control algorithm via anti-aliasing filter and quantization process. After the main algorithms have been executed, digital output signal to adjust dead time of the reclosing relay is sent to this relay. Thus adaptive auto-reclosing, following certain duration times, can be accomplished depending on various faults locations, and this approach is distinctly different from employing a fixed dead time scheme of interrupting duty.

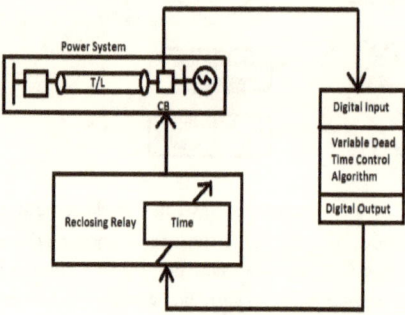

Fig. 6. Variable dead time auto-reclosing scheme

For the simulation purpose, in this work the dead time is

selected from 0.62sec to 1.2secs with an interval of 0.1sec. In this system, the fault duration is 0.5sec. The reason to start the dead time from 0.62 sec is that this time is greater than the sum of the $T_r$as shown in (1) and the fault duration of 0.5 sec. Then based on the voltage index, V_{index}, shown in (6), the ORCT is determined. Tables VI-IX show the indices corresponding to the chosen values of variable dead times. From the Tables VI-IX, the ORCTs are determined and shown in Tables X-XI.

$$V_{index} = \int_0^T |\Delta V| dt \qquad\qquad (6)$$

whereΔV indicates the total voltage deviation of the generators and T is the simulation time of 20.0 sec. The lower the value of the index, the better the system's performance is.

Now using the orcas obtained from the three reclosing methods, the voltage and speed indices as shown in the next section are calculated. based on these two indices, the best reclosing method is selected.

Table IV

Values of Orct for Temporary Fault

Fault Type	Fault point	Reclosing time (sec) with Conventional method	Reclosing time (sec) with Kinetic Energy Method
3LG	F1	1.0	0.889
	F2	1.0	0.87
	F3	1.0	0.873
1LG	F1	1.0	0.81
	F2	1.0	0.85
	F3	1.0	0.848

Table V

Values of Orct for Permanent Faults

Fault Type	Fault point	Reclosing time (sec) with Conventional method	Reclosing time (sec) with Kinetic Energy Method
3LG	F1	1.0	0.889
	F2	1.0	0.871
	F3	1.0	0.873
1LG	F1	1.0	0.81
	F2	1.0	0.8
	F3	1.0	0.729

7.4. Simulation Results and Discussions

In this work, simulations are performed through Matlab/Simulink. In order to demonstrate the effectiveness of the reclosing techniques, simulations have been carried out considering both balanced (3LG: three-phase-to-ground) and unbalanced (1LG: single-phase-to-ground) temporary and permanent faults. F1, F2 and F3 are considered as the fault locations in the system model of Figure 1. The simulation time and time step are considered as 20s and 50 µs, respectively.

It is assumed that the fault occurs at 0.1sec, the circuit breaker opens at 0.2sec, and the circuit breaker recloses at 1.0 sec in case of the conventional reclosing. However, in case of the mentioned 3 reclosing techniques, the circuit breaker recloses at the ORCT values as shown in Tables II, III and IV. In case of permanent faults, the circuit breakers reopen after 0.1sec of the reclosing instances.

The performance of each reclosing method is evaluated in terms of the voltage index already shown in (6), and also in terms of speed index, S_{index}, shown in (7) below. The lower the value of the indices, the better the system's performance is.

$$S_{index} = \int_0^T |\Delta\omega| dt \qquad (7)$$

where $\Delta\omega$ indicates the total speed deviation of the generators and T is the simulation time of 20.0 sec.

137

Table VI

Values of Vindex for temporary 3lg faults

Reclosing time (sec)	3LG		
	F1	F2	F3
	V index	V index	V index
0.62	1.052	0.7735	0.8416
0.72	1.052	0.7734	0.8416
0.82	1.052	0.7734	0.8416
0.92	1.051	0.8416	0.7733
1.02	1.051	0.8415	0.7733
1.12	1.051	0.7732	0.8415
1.2	1.051	0.7732	0.8415

Table VII

values of Vindex for temporary 1lg faults

Reclosing time(sec)	1LG		
	F1	F2	F3
	V index	V index	V index
0.62	0.7756	0.775	0.7751
0.72	0.7756	0.775	0.7751
0.82	0.7756	0.775	0.7751
0.92	0.7756	0.7749	0.7751
1.02	0.7756	0.7749	0.7751
1.12	0.7756	0.7749	0.7751
1.2	0.7757	0.7749	0.7751

Table VIII	Table IX
Values of Vindex for Permanent 3lg faults	Values of Vindex for Permanent 1lg faults

Table VIII

Values of Vindex for Permanent 3lg faults

Reclosing time(sec)	3LG		
	F1 V index	F2 V index	F3 V index
0.62	1.146	0.8416	0.8705
0.72	1.143	0.8419	0.8713
0.82	1.142	0.8418	0.8715
0.92	1.14	0.8421	0.8715
1.02	1.138	0.842	0.8713
1.12	1.136	0.8418	0.8711
1.2	1.137	0.8417	0.8708

Table IX

Values of Vindex for Permanent 1lg faults

Reclosing time(sec)	1LG		
	F1 V index	F2 V index	F3 V index
0.62	0.7900	0.796	0.7824
0.72	0.7901	0.796	0.7824
0.82	0.7902	0.7961	0.7825
0.92	0.7903	0.7961	0.7825
1.02	0.7904	0.7961	0.7826
1.12	0.7905	0.7962	0.7827
1.2	0.7905	0.7962	0.7828

Table X

Values of Orct for Temporary Faults

Fault Type	Fault point	Reclosing time (sec) with Conventional method	Reclosing time (sec) with Variable Dead Time Method
3LG	F1	1.0	0.92
	F2	1.0	0.72
	F3	1.0	0.92
1LG	F1	1.0	0.92
	F2	1.0	0.72
	F3	1.0	0.72

Table XI

Values of Orct for Permanent Faults

Fault Type	Fault point	Reclosing time (sec) with Conventional method	Reclosing time (sec) with Variable Dead Time Method
3LG	F1	1.0	1.12
	F2	1.0	0.62
	F3	1.0	0.62
1LG	F1	1.0	0.62
	F2	1.0	0.62
	F3	1.0	0.72

Table XII

Values of Vindex and Sindex of Temporary 3lg and 1lg Fault for Conventional and Total Load Angle Reclosing Method

Fault Type	Fault point	Conventional Reclosing based Indices (pu.sec)		Total Load angle Reclosing based Indices (pu.sec)	
		Vindex	Sindex	Vindex	Sindex
3LG	F1	1.057	0.013	1.051	0.0136
	F2	0.7734	0.0019	0.7729	0.0032
	F3	0.8419	0.0077	0.8415	0.0078
1LG	F1	0.7756	0.0007	0.7757	0.0013
	F2	0.7749	0.0005	0.7748	0.0007
	F3	0.7751	0.0017	0.7751	0.0018

7.5. Performance of ORCT Methods and Overall Comparison among Them

Tables XII-XVII show the values of voltage and speed indices for the conventional and optimal reclosing methods in case of both temporary and permanent 3LG and 1LG faults at points F1, F2 and F3. From the indices in Tables XII-XVII, it is clear that any of the mentioned three reclosing methods can perform better than the conventional reclosing method. However, if the indices among the three methods are compared, then it is seen that the total kinetic energy optimal reclosing technique performs the best. Therefore, in the following subsection, the voltage and speed responses are shown for the total kinetic energy based reclosing method and compared those with the conventional reclosing method.

Table XIV

Values of Vindex, and Sindex of Temporary 3lg and 1lg Fault for Conventional and Total Kinetic Energy Reclosing Method

Fault Type	Fault point	Conventional Reclosing based Indices (pu.sec)		Total Kinetic Energy Reclosing based Indices (pu.sec)	
		V_{index}	S_{index}	V_{index}	S_{index}
3LG	F1	1.137	0.0221	1.14	0.0076
	F2	0.8424	0.0054	0.8421	0.0008
	F3	0.8751	0.0081	0.8715	0.0021
1LG	F1	0.7903	0.0022	0.7902	0.0002
	F2	1.092	0.0011	0.7961	0.0002
	F3	0.7827	0.0016	0.7824	0.0001

Table XIII

Values of Vindex and Sindex of Permanent 3lg and 1lg Fault for Conventional and Total Load Angle Reclosing Method

Fault Type	Fault point	Conventional Reclosing based Indices (pu.sec)		Total Load angle Reclosing based Indices (pu.sec)	
		V_{index}	S_{index}	V_{index}	S_{index}
3LG	F1	1.137	0.0221	1.146	0.0358
	F2	0.8424	0.0054	0.8413	0.0084

	F3	0.8751	0.0081	0.8689	0.0161
1LG	F1	0.7903	0.0022	0.7908	0.0028
	F2	1.092	0.0011	0.7963	0.0018
	F3	0.7827	0.0016	0.7832	0.0042

7.6. Generator Speed and Voltage Responses

Figs.7-8 show the voltage and speed responses for temporary and permanent 3LG faults at points F1 in case of the total kinetic energy based reclosing and conventional reclosing methods. From the responses it is evident that the total kinetic energy based method performs better than the conventional reclosing method.

Table XV

Values of Vindex and Sindex of Permanent 3lg and 1lg Fault for
Conventional and Total Kinetic Energy Reclosing Method

Fault Type	Fault point	Conventional Reclosing based Indices (pu.sec)		Total Kinetic Energy Reclosing based Indices (pu.sec)	
		V_{index}	S_{index}	V_{index}	S_{index}
3LG	F1	1.057	0.013	1.051	0.0179
	F2	0.7734	0.0019	0.7734	0.0030
	F3	0.8419	0.0077	0.8416	0.0078
1LG	F1	0.7756	0.0007	0.7756	0.0013
	F2	0.7749	0.0005	0.775	0.0010
	F3	0.7751	0.0017	0.7751	0.0017

Table Xvi

Values of Vindex and Sindex of Temporary3lg and 1lg Fault For F1, F2 and F3 Fault Point by Variable Dead Time Method

Fault Type Temporary	Fault point	Conventional Reclosing Indices		Variable Dead Time Reclosing Indices	
		V_{index}	S_{index}	V_{index}	S_{index}
3LG	F1	1.057	0.013	1.051	0.0172
	F2	0.7734	0.0019	0.7734	0.0034
	F3	0.8419	0.0077	0.7734	0.0034
1LG	F1	0.7756	0.0007	0.7756	0.0011
	F2	0.7749	0.0005	0.775	0.0009
	F3	0.7751	0.0017	0.7751	0.0017

Table Xvii

Values of Vindex and Sindex of Permanent 3lg and 1lg Fault For F1, F2 And F3 Fault Point By Variable Dead Time Method

Fault Type	Fault point	Conventional Reclosing Indices		Variable Dead Time Reclosing Indices	
		V_{index}	S_{index}	V_{index}	S_{index}
3LG	F1	1.137	0.0221	1.136	0.0185
	F2	0.8424	0.0054	0.8416	0.0043
	F3	0.8751	0.0081	0.8705	0.0054
1LG	F1	0.7903	0.0022	0.79	0.0011
	F2	1.092	0.0011	0.796	0.0005
	F3	0.7827	0.0016	0.7824	0.0001

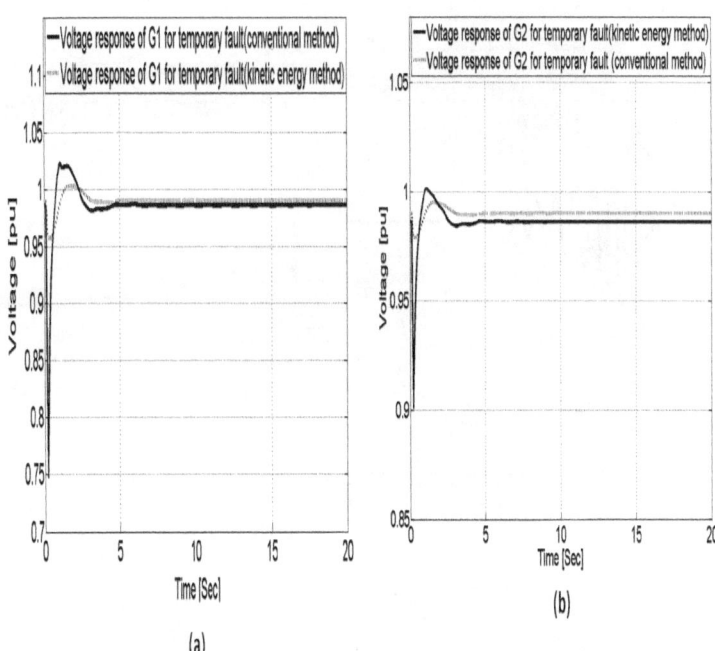

(a)

(b)

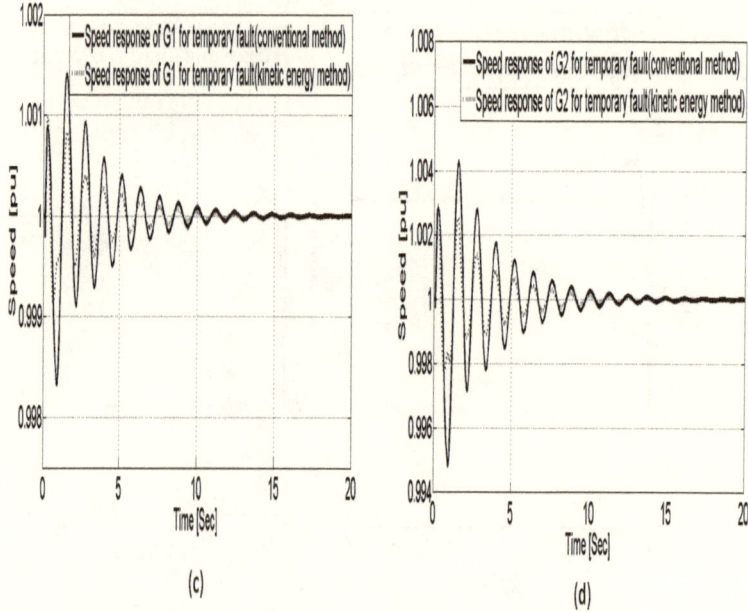

(c)   (d)

Fig.7. Voltage and speed responses for 3LG temporary faults in case of total kinetic energy based optimal reclosing and conventional reclosing methods at point F1.

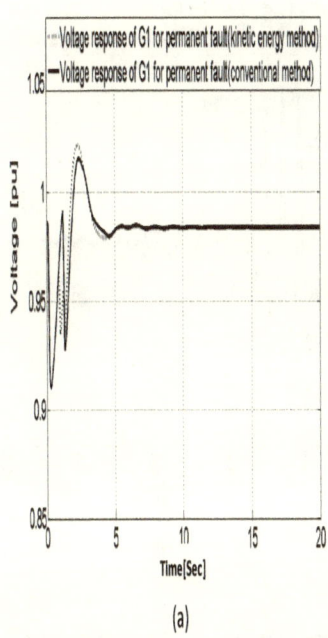

(a)

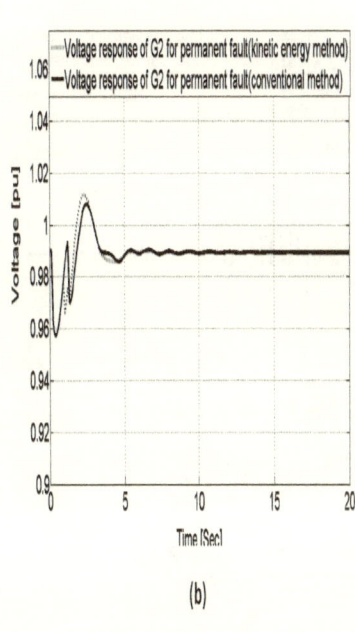

(b)

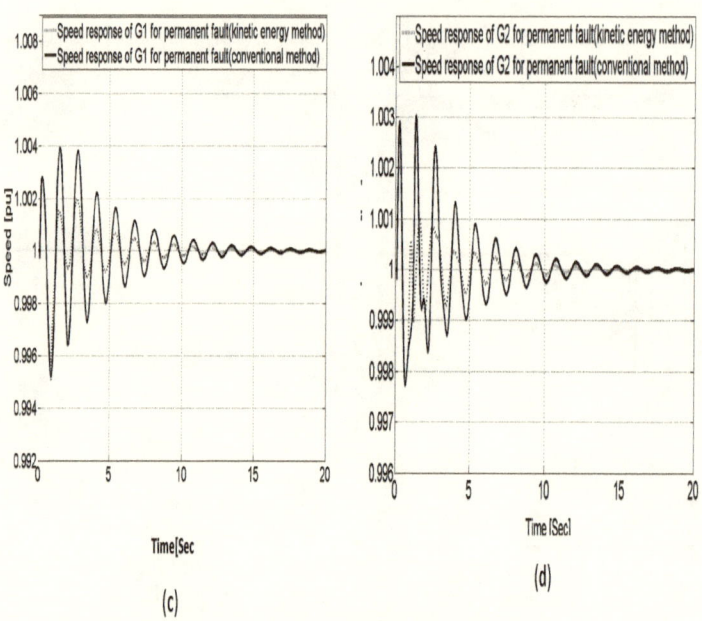

(c)

(d)

Fig.8. Voltage and speed responses for 3LG permanent faults in case of total kinetic energy based optimal reclosing and conventional reclosing methods at point F1.

7.7. Practicality of the Optimal Reclosing Methods

One question might arise here regarding the practical implementation of the optimal reclosing methods. However, it is hoped that the mentioned reclosing methods can be implemented in real time. The required signals needed to be captured for the total kinetic energy based method, total load angle based method, and variable dead time based reclosing methods are speed, load angle, and voltage, respectively. The online measurement of the speed, load angle, and voltage of different generators located at different locations and then calculation of the total kinetic energy, total load angle, etc., can be done through the global positioning system (GPS) [14-17].

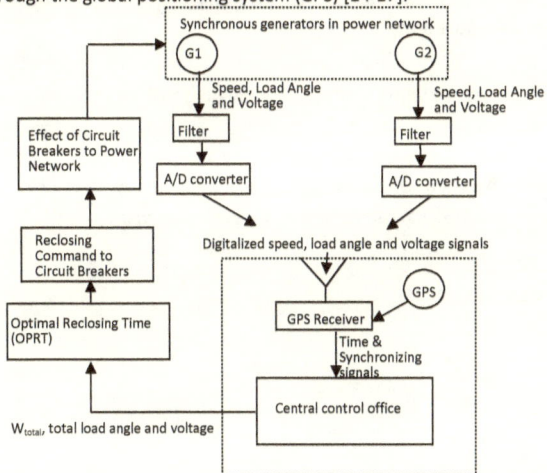

Fig. 9. Closed-loop control system of optimal reclosing methods.

Fig. 9 shows the functional block diagram of GPS, where the GPS receiver receives the digitalized speed/angle/voltage signals collected from the generators, and makes them both time and phase synchronized. Using the computers, the central control office then can determine the ORCT easily.

7.8. Conclusion

This chapter provides a comparative analysis of different types of reclosing techniques, such as the total kinetic energy method, the variable dead time control method, and the total load angle method. The performance of these three methods is compared with that of the conventional reclosing method. Both balanced and unbalanced temporary and permanent faults at different locations in the power system model are considered. From the simulation results, the following points are noteworthy.

a) Any of the mentioned three optimal reclosing methods performs better than the conventional reclosing method.

b) The total kinetic energy based reclosing technique performs the best. Therefore, the total kinetic energy based reclosing technique can be considered as an effective method of power quality enhancement in a multi-machine power system.

7.9. References

1. Sunil S Rao, *Switchgear Protection and Power System*. Khanna Publishers, 2010.

2. P. Kundur, *Power System Stability and Control*. McGraw-Hill, Inc., 1994.

3. B. H. Zhang , Z. Q. Bo, Y. Z. Ge, R. K. Aggarwal, and A. T. Johns, "The design and application of an optimal reclosure technique for transmission lines,"*1995North American Power Symposium,* Montana State University, Bozeman, MT, USA,pp. 415-421, 1995.

4. S. Ghosh and M. H. Ali, "Load Angle Based Optimal Reclosing Technique of Circuit Breakers for Power Quality Enhancement", *Paper ID: 55, Proceedings of the IEEE Southeast Con 2013*, Florida, USA, and April 4-7, 2013.

5. M. Yagami, T. Murata, and J. Tamura, "An analysis of optimal reclosing for enhancement of transient stability", *Electrical Engineering in Japan*, vol.147, no.3, pp.32-39, 2004.

6. S.-P. Ahn, C.-H. Kim, R. K. Aggarwal and A. T. Johns, "An alternative approach to adaptive single pole auto-reclosing in high voltage transmission systems based on variable dead time control*," IEEE Trans. Power Delivery*, vol. 16, no. 4, Oct. 2001, pp. 676-686

7. I.-D. Kim, H.-S. Cho, J.-K. Park, "A variable dead time circuit breaker auto-reclosing scheme using artificial neural networks," *Electrical Power and Energy Systems*, vol. 21, pp. 269-277, 1999.

8. H. –C. Seo and C. –H. Kim, "An adaptive reclosing algorithm considering distributed generation," *International Journal of Control, Automation, and Systems*, vol. 6, no. 5, pp. 651-659, Oct. 2008.

9. Z.Q. Bo, R.K. Aggarwal, A. T. Johns, B. H. Zhang and Y.Z. Ge, "New concept in transmission line reclosure using high- frequency fault

transients," *Proc. Inst. Elect. Eng., Gen., Transm.,Distri.,*, vol.144, no. 4, pp.351-356, Jul. 1997.

10. Y. Yuchun, Z. Baohui and W. Qingfang, "A method for capturing optimal reclosing time of Transient Fault," *In Proc. IEEE International Conf. on Power Sys. Technology, POWERCON 1998*, vol. 2, pp. 1138-1142.

11. B. H. Zhang, Y. C. Yuan, Z. Chen, and Z. Q. Bo, "Computation of optimal reclosure time for transmission lines," *IEEE Trans. Power Sys.*, vol. 17, no. 3,pp. 670-675, August 2002.

12. R. K. Aggarwal, A. T. Johns, Y. H. Song, R. W. Dunn, and D. S. Fitton, "Neural-network based adaptive single-pole autoreclosure technique for EHV transmission systems," *IEE Proc.-Gener. Transm. Distrib.,*vol. 141, no. 2, pp. 155-160, March 1994.

13. M. H. Ali, T. Murata, and J. Tamura, "Effect of Coordination of Optimal Reclosing and Fuzzy Controlled Braking Resistor on Transient Stability During Unsuccessful Reclosing," *IEEE Trans. Power Sys.*, vol. 21, no. 3, pp. 1321-1329 , August 2006.

14. Working Group H-8 of the relay Communications Subcommittee of the IEEE Power System Relaying Committee, "IEEE standard for synchrophasors for power systems," *IEEE Trans. Power Del.*, vol. 13, no. 1, pp. 73-77, Jan 1998.

15. H.Y. Li, E.P. Southern, P.A. Crossley, S. Potts, S.D.A. Pckering, B.R.J. Caunce, and G.C. Weller, "A new type of differential feeder protection relay using the global positioning system for data synchronization ," *IEEE Trans. Power Del.*, vol. 12, no. 3, pp. 73-77, Jul 1997.

16. Working Group H-7 of the Relaying Channels Subcommittee of the IEEE Power System Relaying Committee, " Synchronized sampling and phasor measurements for relaying and control," *IEEE Trans. Power Del.*, vol. 9, no. 1, pp. 442-452, Jan 1994.

17. R.O. Burnett Jr., M.M. Butts, T.W. Cease, V. Centeno, G. Michale, R.J. Murphy and A.G. Phadke," Synchronized phasor measurements of a power system event," *IEEE Trans. Power Del.*, vol. 9, no. 3, pp. 1643-1650, Aug 1994.

Chapter 8

Multi-Objective Reconfiguration in Electric Distribution Systems

Esmaeel Rok Rok, Mahmoud Reza Shakarami, Hajar Bagheri Tolabi
and Rahil Hossieni

8.1. Introduction

Distribution system is an interface between the consumers and transmission networks. Due to the advantages such as lower short circuit current and easier protection coordination, the distribution systems are generally utilized in radial structure. The radial structure can increase the active power losses, voltage drops at the load points, etc. Electrical power distribution systems have two types of tie and sectionalizing switches, whose statuses determine the configuration of distribution network. By changing the switches states and transition of sections between feeders during operation, the construction of distribution network will change. This configuration change of distribution system is known as reconfiguration and performed for different objectives such as reduction of losses, improving the voltage profiles, load balancing and etc. Many researches in the literatures have presented several methods for the optimal reconfiguration of the distribution networks with different objectives. Reconfiguration of distribution network for loss reduction was first proposed by Merlin and Back in 1975. They have used a branch and bound optimization method to determine the configuration that has the minimum total loss. In this method, all switches are first closed to establish a meshed

155

configuration. The switches are then opened successively to achieve the radial configuration. After that, many algorithms have been developed for reconfiguration of distribution system with different aims. Goswami and Basu presented a heuristic algorithm for reconfiguration which is determined using a power flow program. Gomes et al. reported a heuristic algorithm for the large distribution systems that begins in a meshed configuration with all switches closed. Switches will be opened one by one according to the calculation of the minimum system loss using a power flow program. A new path to node based modeling and its application to reconfiguration of distribution system has been proposed by Ramoset and Exposito in 2005. Schmidtet al. has introduced a method for loss minimization based on the standard Newton technique. Zhou et al. have presented two reconfiguration algorithms for service restoration and load balancing in distribution systems. They used the combination of heuristic rules and fuzzy logics in optimization purposes to solve the reconfiguration problem. An optimization technique to determine the network structure with minimum energy losses for a given period has proposed by Taleski and Rajicic. The application aspects of optimal distribution system reconfiguration have been considered by Borozan and Rajakovic. To determine the switching operation, an algorithm of network reconfiguration using voltage, ohmic, and decision indexes has been presented by Lin and Chin. Jeon, Augugliaro and their colleagues used artificial-intelligence in a minimum loss reconfiguration. Nara et al. Have solved distribution reconfiguration problem for minimum loss using Genetic Algorithm (GA). A fuzzy multi-objective approach was offered by Das for optimizing distribution network configuration which four objectives load balancing among the feeders, real power loss, deviation of nodes voltage, and branch current constraint violation are modeled and results obtained are encouraging, but criteria for selecting a membership function for each objective are not provided. Harmony

156

Search Algorithm (HSA) is used for optimal network reconfiguration of large distribution system by Rao et al. [16]. In this study Imperialist Competitive Algorithm (ICA) and Graph theory are implemented in Matlab software to solve a multi-objective distribution system reconfiguration problem. The proposed technique is tested on a typical distribution system and the results are presented in numerical studies section. As well as, to investigate the efficiency of proposed method compare with other researches, it is accomplished a comparison between proposed technique and other method results.

8.2 Imperialist Competitive Algorithm

Imperialist competitive algorithm is a computational method that is used to solve the different types of optimization problems. This algorithm is based on assimilation and competition policies. Imperialist countries are trying to attract other colonies to increase power of their governments and eventually dominate on the whole world (at this time the issue has been solved).

a. ICA Overall Progress

This algorithm begins with a random initial population which they are called countries. Some of the best elements of the population are selected as imperialists and other are considered as colony countries. Imperialists depending on power which varies inversely with

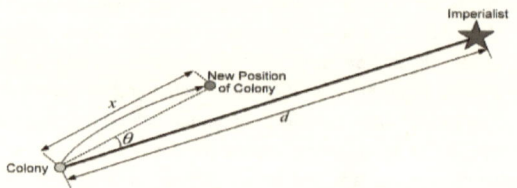

the cost function will attract the colonies by a particular process [19].

Fig 1. Colony movement towards imperialist

Figure1 shows the movement of colonies toward the imperialist. The imperialist countries attract the colonies according to language and culture parameters. In this figure, the distance between the colony and the imperialist is shown by d. x and θ are random numbers with uniform distributions which are defined by equation 1.

$$x \approx U(0, \beta \times d), \theta \approx U(-\gamma, \gamma) \qquad (1)$$

$\beta > 1$ makes different directions to the imperialist. In equation 1, γ is a optional parameter that by increasing it, the search space around the imperialists will be larger and its reduction lead to colonies move nearby the interface vector between the colony and imperialist. In most implementations, $\pi/4$ and 2 values are suitable choices for γ and β parameters respectively [20].

Imperialist competitive algorithm progress is described as following steps:

a. Define an objective function.

b. Generation some random solutions in the search space and create initial empires.

c. Assimilation: colonies move towards imperialists from different directions.

d. Revolution: Random changes occur in the characteristics of some countries.

e. Exchange position between a colony and Imperialist. A colony with a better position than an imperialist has the chance to take the control of empire by replacing the existing imperialist.

f. Imperialistic competition: All imperialists compete to take possession of colonies from each other.

g. Eliminate the powerless empires. Weak empires lose their power gradually and finally, they will be eliminated.

h. Check the stopping criteria: if the stop condition is satisfied, then stop, else go to step c.

8.3 Graph Theory

Each graph is introduced as a binary mixture $G(V,E)$[17], in which the vector V is the set of vertices or nodes in the graph and the vector E is the set of unordered pairs of different vertices that each of them called an edge. The degree of each vertex is determined by the number of edges meeting it. The connected graph is a graph that there is at least one path between each two arbitrary vertices of it. Tree graph is a graph that is connected and there is no cycle in it. In a tree graph, if V is the number of vertices and E is the number of edges, so the following equation is satisfied:

$$V = E - 1 \qquad\qquad (2)$$

The number of cycles in a graph is given by the following equation:

$$cycle = (E + V) - 1 \qquad\qquad (3)$$

A graph with n vertices can be defined by the following square adjacency matrix:

$$A = [a_{ij}]_{n \times n} \qquad\qquad (4)$$

In this matrix, if any two vertices i and $j\,(i \neq j)$ are directly connected together, their corresponding element (a_{ij}) in the matrix is equal to one; otherwise it is equal to zero.

Graph theory can be used to check whether radial structure of distribution system is retained (tree graph) as well as all loads being in service (connected graph).

8.4 Problem Formulation

As mentioned in first section, the reconfiguration of distribution feeders performs with different purposes in the normal operating conditions through the change of closed-open switches statuses.

In the next parts of this section, four different goals have been introduced for the multi-objective reconfiguration problem and each of them will be formulated separately in the form of mathematically models.

A. Minimization of power losses

Distribution network active losses is very high compare with the transmission lines and in a multi-objective reconfiguration problem, reduction of power losses is usually considered as the most important objective function. In this study, power losses objective function is expressed as following mathematically equation:

$$f_1 = \sum_{k=1}^{n} r_k . I_k^{2} \qquad (5)$$

Where:

n : Total number of sections.

r_k : The conductor resistance of section k.

I_k : Current of section k.

B. Minimization of feeder load unbalancing

Feeder load unbalancing is a common problem in the distribution networks. There are different reasons that lead to unbalancing on the feeder load such as heterogeneous and none uniform distribution of single phase subscribers among the three phases of a feeder or random and asynchronous behavior of the single phase consumers.

In this objective function the overall goal is reduction of power imbalance (active and reactive) for all sections and it is formulated as follows:

$$f_i = Variance\left[\frac{S_1}{S_{1max}}, \frac{S_2}{S_{2max}}, ..., \frac{S_k}{S_{kmax}}, ...\frac{S_n}{S_{nmax}}\right]$$ (6)

Where:

S_k : Is the apparent power value flowing through the section k.

S_{kmax} : Is the maximum permissible apparent power value of section k.

C. Deviation of the bus voltage minimization

Since the bus voltage is one of the most important security and service quality indices, basic purpose of this objective function is that the deviation of nodes voltage should be less and it is evaluated as:

$$\min f_d = \max\left|V_k - V_r\right| \quad k = 1,2,...,n_l$$ (7)

Where:

n_l : Total number of buses.

V_k : The real voltage on bus k.

V_r : The rated voltage on bus k.

D. Minimizing the number of switching operations

In distribution system reconfiguration, the optimal structure has to be achieved by changing the open-close status of the switches. The number of switching operations should be as minimal as possible to enhance the life of switchgears, reduce the operating cost and switching transients.

In order to perform the shift from the base configuration to the optimal configuration by minimum switch operations, an efficient switch scheme needs to be developed such that dispensable switch operations can be avoided. Minimizing the number of switch operations can be defined as follows:

$$\min f_s = \sum_{k=1}^{n} |SW_k - SW_{ak}|$$

(8)

Where:

SW_k : The new state of the switch k

SW_{ak} : The original state of the switch k

8.4 Solution Of Reconfiguration Problem Based On Imperialist Competitive Algorithm And Graph Theory

In this study, reconfiguration is used to achieve four minimization targets including power losses, feeder load unbalancing, voltage drop and number of switching operations. The mathematical model of this multi-objective problem is expressed as follows:

$$F = Minimum \left[a_1 . f_l + a_2 . f_i + a_3 . f_d + a_4 . f_s + B \right],$$

$$\sum_{i=1}^{4} a_i = 1 \qquad (9)$$

and the problem constrains are:

$1 : V_{k\min} \leq V_k \leq V_{k\max} \quad k = 1,2,\ldots n_l$

$2 : I_k \leq I_{k\max} \quad k = 1,2,\ldots n$

3: Radial structure of network should be maintained.

4: All available nodes of considered distribution system should be fed.

In the proposed multi-objective function, a_i is a weighting coefficient for each objective function and B is defined as follows:

$$B = C_{mesh(number)} + D_{isolated-loads(number)} \qquad (10)$$

In the equation (10), $C_{mesh(number)}$ and $D_{isolated-loads(number)}$ values are the penalty coefficients for wrong solutions that lead to form rings or isolated loads. By placing parameter B in the multi-objective function, the possible wrong solutions of search space have been deleted gradually and algorithm will converge to the optimal solution with more acceptable speed. 1-4 above mentioned constraints represent the limitation of voltage and current, necessity to maintain the radial structure of distribution system and being all loads in service for reconfiguration problem solving.

Reconfiguration of distribution system processes using the imperialist competitive algorithm and graph theory is described as a bellow flowchart: (The distribution network power flow is performed using the method presented in [18].)

Step1. Read data of distribution system (bus, load, and branch data). Close all switches at network.

Step 2. Generate initial countries. The countries have been shown by switches that can be open or closed.

Step 3. Run the power flow program [18].Compute the power losses, feeder load unbalancing, voltage deviation and number of switching operation values. Calculate multi -objective function value (F_{best}).

Step 4. Select the imperialist countries states and their relevant colonies.

Step 5. Move the colonies toward their imperialist and exert revolution.

Step 6. Calculate the cost function for all countries.

Step 7. If there is a colony in an empire which has lower cost than that of imperialist, so displace between imperialist and colony.

Step 8. Select the weakest colony of the weakest empire and give to the empire that has the most likelihood to process it.

Step 9. If there is an empire with no colony, eliminate it.

Step10. Run the power flow program and calculate multi -objective function value for the new structure (F_{new}).

Step 11. Select lesser multi-objective function between F_{best} and F_{new} and replace it by F_{best} .

Step 12. If the stop condition is not satisfied, go the step 5.

Step 13. Use the graph theory to check the radial structure and feeding all loads of distribution system by the following method:

Find the adjacency matrix according to the obtained ICA solution in the step 12.This matrix indicates the connection between buses of the feeder and is named A1.The matrix A2 will be obtained by eliminating the repeated elements of A1. None zero elements number of the matrix A2 is compared by the bus numbers, so one of the two following states will be achieved: these two numbers are not equal together (state 1) or they are equal with together (state 2).

-State 1: it means that the simulated distribution system graph is not a connection graph and there are some isolated loads. The numbers of isolated loads are equal to the differences between the one number elements of the matrix A2 and the bus numbers.

-State 2: by eliminating the first column of the A1 matrix, if a row with zeros elements is produced, so there are some isolated buses. These buses are those buses which are connected to the feeders (only one feeder assumed in this study). If there are no isolated loads and graph is connected graph, then the numbers of cycles are calculated by equation (3). If the cycle numbers is equal to zero, it is a tree graph and the distribution system is radial and all loads are in service, else multi-objective function should be penalized using new *B* factor and process repeat again from step5. The checking procedure using graph theory is shown as a flowchart in figure2.

Step 14. Final suggested structure.

8.5 Numerical Studies

To evaluate the performance of the proposed model, a sample distribution feeder is tested that its single line diagram is shown in the figure3. It is a 12.66 kV network and consists of one feeder, 33 buses, 32 normally closed switches (sectionalizing switches), and 5 normally open switches (tie switches). Electrical information of the sample distribution system is presented in the table1. It is assumed that all sections have switches. Two different experiments have been done to investigate the efficiency of the proposed method:

First test: In this test, unequal weighting coefficients are considered, $a_1 = 0.4$, $a_2 = 0.4$, $a_3 = 0.1$ and $a_4 = 0.1$ Second test: In this test, all four goals of the scheme are assumed by similar coefficients. In other words, $a_1 = a_2 = a_3 = a_4 = 0.25$.

The obtained results are presented in the table2. As can be seen in this table, the amount of loss reduction is significant for both tests. This reduction is 33.156 % and 30.843% for first and second tests respectively. The feeder load unbalancing has been improved about 36.134% for first and 26.050% for second test. Voltage deviation is reduced for both tests, although further improvement has been seen for second test. The obtained objective function values for the number of switching operations are 8 and 6 in first and second tests respectively. The convergence process of f_l and f_d objective functions have been shown in figures 4and 5 for both tests. The results of the proposed algorithm have been compared with other works in table3. The performance of ICA is satisfactory using the parameter values have been shown in the table4.

The results show that weighting coefficients values changes lead to different solutions. It is observed that the proposed model is a sensitive multi-objective function and able to adapt itself with the new conditions of the project.

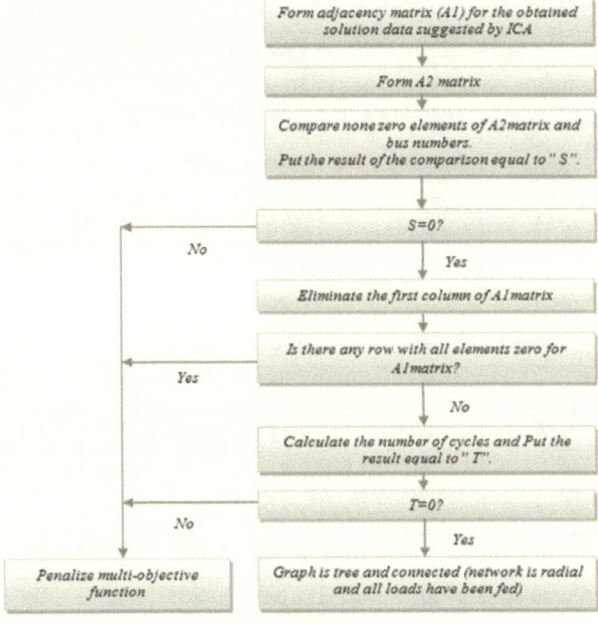

Fig 2: Checking the radial structure and being all loads in service using graph theory.

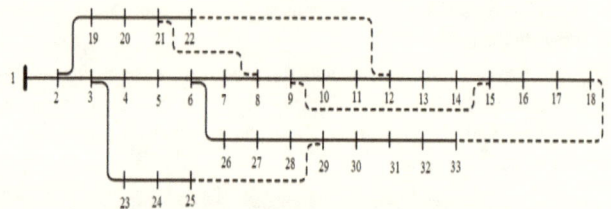

Fig 3: Initial single line diagram of studied feeder.

Table1

Numerical information of studied feeder

Section number	Beginning	End	R(Ohm)	X(Ohm)	P(KW)	Q(KVAR)
1	1	2	0.0922	0.0470	100	60
2	2	3	0. 4930	0.2511	90	40
3	3	4	0. 3660	0.1864	120	80
4	4	5	0. 3811	0.1941	60	30
5	5	6	0. 8190	0.7070	60	20
6	6	7	0. 1842	0.6188	200	100
7	7	8	0. 7114	0.2351	200	100
8	8	9	1. 0300	0.7400	60	20
9	9	10	1. 0440	0.7400	60	20
10	10	11	0. 1966	0.0650	45	30
11	11	12	0. 3744	0.1238	60	35
12	12	13	1. 4680	1.0550	60	35

13	13	14	0.5416	0.7129	120	80
14	14	15	0.5910	0.5260	60	10
15	15	16	0.7463	0.5450	60	20
16	16	17	1.2890	1.7210	60	20
17	17	18	0.7320	0.5740	90	40
18	2	19	0.1640	0.1565	90	40
19	19	20	1.5042	1.3554	90	40
20	20	21	0.4095	0.4784	90	40
21	21	22	0.7089	0.9373	90	40
22	3	23	0.4512	0.3083	90	40
23	23	24	0.8980	0.7091	420	200
24	24	25	0.8960	0.7011	420	200
25	6	26	0.2030	0.1034	60	25
26	26	27	0.2842	0.1447	60	25
27	27	28	1.0590	0.9337	60	20
28	28	29	0.8042	0.7006	120	70
29	29	30	0.5075	0.2585	200	600
30	30	31	0.9742	0.9630	150	70
31	31	32	0.3105	0.3619	210	100
32	32	33	0.3410	0.5320	60	40
33*	21	8	2.0000	2.0000	-	-
34*	9	15	2.0000	2.0000	-	-

35*	12	22	2.0000	0.5000	-	-
36*	18	33	0.5000	0.5000	-	-
37*	25	29	0.5000	0.5000	-	-
*Open branches						

Table 2
The obtained results for both tests

proposed method (second test)	proposed method (first test)	Initial state	Parameters
140.201	135.512	202.730	Objective function value of (KW) f_l
30.843	33.156	-	Real power loss (%) reduction
6	8	-	Objective function value of f_s
5	5	5	Number of tie switches
37-12-17-10-7	32-24-13-9-7	33-34-35-36-37	Section of tie switches
0.088	0.076	0.119	Objective function value of (%) f_i
26.050	36.134	-	Load unbalancing

			improvement(%)
0.0580	0.0618	0.0962	Objective function value of f_d
0.9419	0.9382	0.9038	Worst voltage in p.u.

Table3

Comparison between the obtained results by the proposed method and other researches for optimizing the total active power losses

Method	Tie switches	Power loss(KW)
Original configuration[23]	33-34-35-36-37	202.730
Shirmohammadi and Hong (1989)[22]	7-10- 14- 32- 37	140.26
Goswami and Basu (1992)[1]	7-9- 14- 32- 37	139.53
GA (Nara,1992) [14]	33-9-34- 28- 36	140.6
Vanderson Gomes et al. (2005)[21]	7- 9- 14- 32- 37	139.53
proposed method (first test)	32-24-13-9-7	135.512

Table4
ICA selected parameters

Number of population	37
Number of imperialists	5
Number of colonies	32
Revolution rate	0.1
Iterations	50

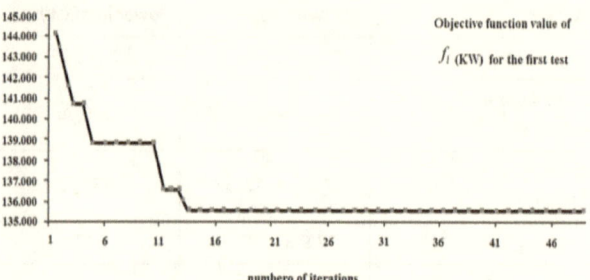

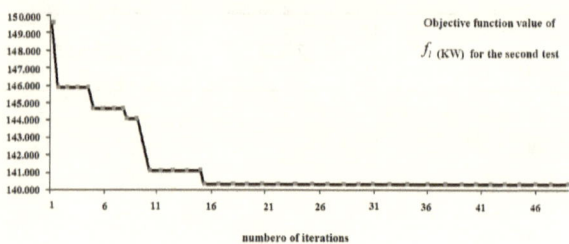

Figure4: f_1 objective function convergence process

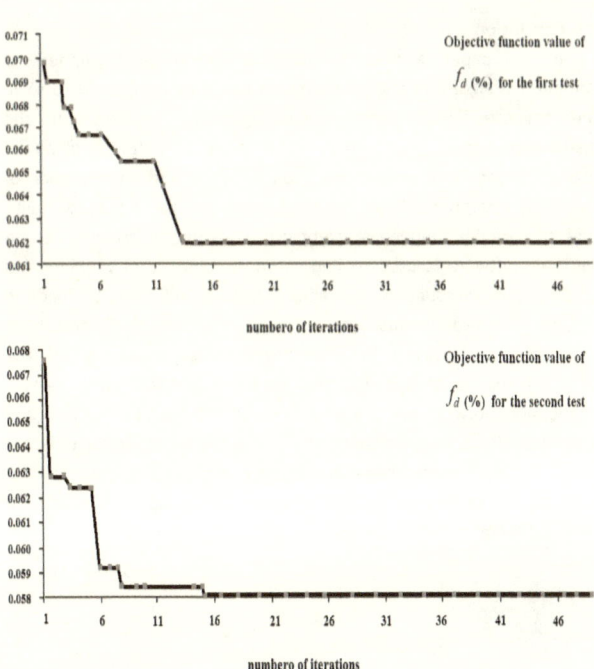

Figure5: f_d **objective function convergence process**

8.6 Conclusion

In this paper, a new technique of combining the imperialist competitive algorithm and graph theory has been introduced to solve the proposed multi-objective reconfiguration problem for the distribution system. Four targets including the minimization of power losses, feeder load unbalancing, voltage drop and the number of switching operations have been assumed in the proposed multi-objective model. Imperialist competitive algorithm implemented in MATLAB software is used to find the best solutions of problem and graph theory investigates the radial structure and being all loads in service. Two experiments with different weighting coefficients have been done to investigate the efficiency of the proposed method. The obtained results of applying the proposed method on a sample distribution network are presented in numerical studies section and prove the efficient performance of the suggested method for different states. Among the other benefits of this method is that operator can use each of the targets separately or establish the optimal interaction between them.

References

1. S.K.Goswami and S.K. Basu. A new algorithm for the reconfiguration of distribution feeders for loss minimization. IEEE Trans. Power Del. vol.7, n.3, 1992, pp.1484 –1491.

2. F. V. Gomes and S. Carneiro, Jr. A new heuristic reconfiguration algorithm for large distribution systems. IEEE Trans. Power Syst. vol.20, n.3, 2005, pp.1373 –1378.

3. E. R. Ramos and A. G. Exposito. Path-based distribution network modeling: Application to reconfiguration for loss reduction. IEEETrans. Power Syst. vol.20 n.2, 2005, pp. 556 –564.

4. H. P. Schmidt and N. Kagan. Fast reconfiguration of distribution systems considering loss minimization. IEEE Trans. Power Syst. vol. 20 n.3, 2005, pp.1311–1319.

5. Lopez, E. and Opaso, h.online reconfiguration considering variability demand: Applications to real networks.IEEE Transactions on Power Systems. vol.19 n.1, 2004, pp.549–553.

6. V. Borozan, D. Rajicic, and R. Ackovski. Improved Method for Loss Minimization in Distribution Networks. IEEE Transaction on Power System. vol. 10 n.3, 1995, pp.1420–1425.

7. A. Merlin and H. Back. Search for a minimal-loss operating spanning tree configuration in an urban power distribution system. in Proc. 5th Power System Computation Conf., Cambridge, pp.1–18,1975,U.K.

8. Q. Zhou, D. Shirmohammadi and W. H. E. Liu. Distribution Feeder Reconfiguration for Service Restoration and Load Balancing. IEEE Trans. Power Syst. vol.12, n.2, 1997, pp. 724–729.

9. R. Taleski and D. Rajicic. Distribution Network Reconfiguration for Energy Loss Reduction. IEEE Transaction on Power System. vol.12, n.1, 1997, pp. 398–406.

10. V. Borozan and N. Rajakovic. Application Assessments of Distribution Network Minimum Loss Reconfiguration. IEEE Transaction on Power Delivery. vol.12 n.4, 1997, pp.1786–1792.

11. W. M. Lin and H. C. Chin. A New Approach for Distribution Feeder Reconfiguration for Loss Reduction and Service Restoration. IEEE Transaction on Power Delivery. vol.13 n.3, 1998, pp.870–875.

12. Y. J. Jeon, J. C. Kim, J. O. Kim, J. R. Shin, and K. Y. Lee. An Efficient Simulated Annealing Algorithm for Network Reconfiguration in Large-Scale Distribution Systems. IEEE Transaction on Power Delivery. vol.17 n.4, 2002, pp.1070–1078.

13. A. Augugliaro, L. Dusonchet, M. Ippolito, and E. R. Sanseverino. Minimum Losses Reconfiguration of MV Distribution Networks Through Local Control of Tie-Switches. IEEE Transaction on Power Delivery. vol.18 n.3, 2003,pp. 762–771.

14. K. Nara, A. Shiose, M. Kitagawa, and T. Ishihara. Implementation of Genetic Algorithm for Distribution System Loss Minimum Reconfiguration. IEEE Transaction on Power Delivery. vol. 7 n.3, 1992, pp.1044–1051.

15. D. Das. A Fuzzy Multi-Objective Approach for Network Reconfiguration of Distribution Systems. IEEE Transaction on Power Delivery. vol. 21, n.1, 2006, pp.202–209.

16. R.S. Rao, S.V.L. Narasimham, M.R. Raju, and A.S. Rao. Optimal Network Reconfiguration of Large-Scale Distribution System Using Harmony Search Algorithm. IEEE Transactions on Power Systems. vol. 26 n.3, 2011, pp.1080-1088.

17. R. Diestel. Graph theory. Springer-Verlag Heidelberg, 2005,New York.

18. S.Ghosh and D.Das. Method for load-flow solution of radial distribution networks. IEE Proc. Gener. Transm. Distrib.Vol.146 n.6, 1999, pp.641 – 648.

19. Atashpaz Gargari. E and Lucas. C. Imperialist Competitive Algorithm: An algorithm for optimization inspired by imperialistic competition. IEEE Congress on Evolutionary Computation, vol.7, 2007, pp. 4661–4666.

20. E.Atashpaz Gargari and C. Lucas. Designing an optimal PID controller using Colonial Competitive Algorithm. First Iranian Joint Congeress on Intelligent and Fuzzy Systems, September 2007, Mashhad-Iran.

21. Vanderson Gomes. F., Carneiro.S., Pereira. J. L. R., Garcia Mpvpan, and Ramos Araujo, L. A new heuristic reconfiguration algorithm for large distribution systems. IEEE Transactions on Power Systems, vol. 20 n.3, 2005, pp.1373–1378.

22. Shirmohammadi. D. and Hong, H. W. Reconfiguration of electric distribution networks for resistive line loss reduction. IEEE Transactions on Power Systems, vol.4 n.1, 1989, pp.1492–1498.

23. Baran, M. E. and Wu, F. F. (1989). Network reconfiguration in distribution systems for loss reduction and load balancing. IEEE Transactions on Power Delivery, vol.4 n.2, 1989, pp.1401–1407.